Bonnie Mage Morrison

THE URBAN ECOSYSTEM:
A Holistic Approach

COMMUNITY DEVELOPMENT SERIES

Series Editor: Richard P. Dober, AIP

Volumes Published and in Preparation

TERRAIN ANALYSIS: A Guide to Site Selection Using Aerial Photographic Interpretation/Douglas S. Way

URBAN ENVIRONMENTS AND HUMAN BEHAVIOR: An Annotated Bibliography/Edited by Gwen Bell, Edwina Randall, and Judith E. R. Roeder

ENVIRONMENTAL DESIGN RESEARCH, VOL. I: Selected Papers/Edited by Wolfgang F. E. Preiser

ENVIRONMENTAL DESIGN RESEARCH, VOL. II: Symposia and Workshops/Edited by Wolfgang F. E. Preiser

APPLYING THE SYSTEMS APPROACH TO URBAN DEVELOPMENT/Jack LaPatra

DESIGNING FOR HUMAN BEHAVIOR: Architecture and the Behavioral Sciences/Edited by Jon Lang, Charles Burnette, Walter Moleski, and David Vachon

ALTERNATIVE LEARNING ENVIRONMENTS/Edited by Gary J. Coates

BEHAVIORAL RESEARCH METHODS IN ENVIRONMENTAL DESIGN/Edited by William Michelson

PLANNING URBAN ENVIRONMENT/Edited by Melville Branch

STRATEGY FOR NEW COMMUNITY DEVELOPMENT IN THE UNITED STATES/Edited by Gideon Golany

LANDSCAPE ASSESSMENT: Values, Perceptions, and Resources/Edited by Ervin H. Zube, Robert O. Brush and Julius Gy. Fabos

Community Development Series

THE URBAN ECOSYSTEM:
A Holistic Approach

THE INSTITUTE OF ECOLOGY
Report of the Urban
Ecosystems Project

Supported by
**RESEARCH APPLIED TO
NATIONAL NEEDS**
National Science Foundation

Edited by
FOREST STEARNS

TOM MONTAG

Artwork by
CHARLES HOLZBOG

 Dowden, Hutchinson & Ross, Inc.
Stroudsburg, Pennsylvania

distributed by
Halsted Press
a division of John Wiley & Sons, Inc.

Library of Congress Cataloging in Publication Data
Main entry under title:

The Urban ecosystem : a holistic approach.

   (Community development series ; v. 14)
  Report of a workshop sponsored by the Institute of Ecology and held at the University of Texas at Austin, Mar. 29-Apr. 4, 1973.
   Includes bibliographical references and index.
   1. Cities and towns--Congresses. 2. Human ecology --Congresses. 3. Cities and towns--Planning--Congresses. I. Stearns, Forest, ed. II. Montag, Tom, ed. III. Institute of Ecology.
HT107.U72   301.36   75-1001
ISBN 0-470-82079-9

Copyright © 1974 by Dowden, Hutchinson & Ross, Inc.
Library of Congress Catalog Number:
ISBN:

75 76 5 4 3 2 1

All rights reserved. No part of this book covered by the copyrights hereon may be reproduced or transmitted in any form or by any means—graphic, electronic, or mechanical, including photocopying, recording, taping or information storage and retrieval systems—without written permission of the publisher.

Manufactured in the United States of America.

Exclusive distributor: HALSTED PRESS,
a division of John Wiley & Sons, Inc.

Contents

PART ONE

SERIES EDITOR'S PREFACE	vii
PREFACE	ix
LIST OF PARTICIPANTS	xiii
INTRODUCTION	3
MAJOR RECOMMENDATIONS	6
Section I: Goals: Values and Jurisdictions	6
Section II: Components: Population, Physical Structure and Resources	8
Section III: Processes: Interventions, Indicators and Models	11
Section IV: Case Studies	13
TOWARDS THE URBAN ECOSYSTEM	15
Cities	17
Ecosystems and Urban Systems	24
Towards the Urban Ecosystem	30
References	31

PART TWO

I. GOALS: VALUES AND JURISDICTIONS	35
Goals	37
Changing Human Values and Behavior	41
Institutions	44
Existing Institutions	46
New Institutions	50
Links Between Institutions	55
Ecological Accounting Systems	59
Continuous Evaluation of Developments	60
Summary	62
Recommendations	64
References	66

II. COMPONENTS: POPULATION,
PHYSICAL STRUCTURE,
AND RESOURCES 68
   Determinants of Urban Population Size 70
   Distribution Trends 76
   Non-Human Populations in the City 81
   Physical Structure of the Urban Ecosystem 85
   Function of the Urban Center 92
   Resources: Water, Energy and Materials Flows 98
   A Conceptual View of the Urban System 104
   Summary 107
   Recommendations 109
   References 111
III. PROCESSES: INTERVENTIONS,
INDICATORS AND MODELS 115
   Altering Relationships Between Urban Structure and Function 117
   Intervening in the Materials and Energy System 120
   Evaluating Alternative Interventions 127
   Alternative Strategies 131
   Indicators of Urban Ecosystem Health 138
   Indices of Critical Processes and States 144
   Diagramming Urban Processes 151
   The Operational Framework of Social-Environmental Accounting 153
   Criteria for Selecting Indicators 156
   Problems of Aggregation and Disaggregation 157
   Indicators in Public Policy 159
   Indicators of Stability and Resiliency 161
   System Models 163
   Summary 173
   Recommendations 174
   References 177
IV. CASE STUDIES 180
   Hierarchical Environmental Analysis 182
   The Butte, Montana, Rudat Study 192
   The Minneapolis-St. Paul Study 195
   The Halifax, Nova Scotia, Study 197
   The Waterloo, Ontario, Study 197
   The Knoxville, Tennessee, Study 200
   The Norman, Oklahoma, Model 202
   Summary 202
   Recommendations 203
   References 203
INSTITUTIONAL AFFILIATIONS OF PARTICIPANTS
   INDEX

Series Editor's Preface

The systems approach to community development requires understanding of the relationships between various components of the urban community. Uncovering those relationships is no easy task. To make the relationships visible enough for constructive action to be taken is more difficult. Translating that which seems apparent into legislative programs and executive actions may be the most challenging effort of all. Where does one begin?

An ecological approach to urbanization may help move us towards understanding and wise action in the urban system. The approach has many advantages. Knowledge of important interrelationships can help us avoid isolated solutions to urban problems. If a medical analogy for design decision-making comes readily to mind—diagnosis, prognosis, prescription—well and good. The body politic is unhealthy.

In this book, nearly one hundred bright minds have defined important urban issues from an ecological point of view. They suggest what might be done to respond to urban needs. They attempt to restore some sense of curative order and connectiveness between the natural and the man-made world, so that the space man occupies could beneficially support all life forms.

Richard P. Dober, AIP

Preface

Urban environments of North America seem unable to support healthy human populations at tolerable levels of stress and to provide rich social and cultural opportunities. Technological changes have placed man increasingly at odds with his life support systems. Further, in his recent cultural evolution, man has created urban complexes at odds with his behavioral and perceptual patterns. Man is, it seems, increasingly out of place in the city, yet increasingly dependent upon it.

Central to the problems urban man faces is a general lack of sensitivity to man's biological nature and his relationship to his environment.

The Urban Ecosystem Project of The Institute of Ecology (TIE) was undertaken in late 1971 with the hope that urban problems could be avoided or ameliorated if knowledge of the ecological interrelationships of people and their urban environments were transmitted to those responsible for urban planning and management. The project specified as first steps: the creation of a multi-disciplinary advisory council; the convening of a national worship to identify national and regional urban needs; and the development and stimulation of communication between various disciplines and practitioners.

With guidance and funding from the Division of Environmental Systems and Resources in the National Science Foundation's Research Applied to National Needs (RANN) program the project became active in June of 1972. An advisory council was appointed to develop a workshop format and specify workshop objectives, assist in selection of participants, and designate task group leaders. In addition to the advisory council approximately 90 scholars and practitioners became involved, serving on eight task groups:

—Goals for the Urban Ecosystem
—Physical Structure and Function
—Resources: Water, Energy, Materials
—Population Processes
—Institutional Structures
—Case Studies of Cities and Urbanized Regions
—Indicators of Urban Ecosystem Function and Health
—Systems Capacities, Limits and Intersystem Linkages

Each task group included participants representing disciplines from the natural and social sciences, especially balanced for the particular task at hand. Task group topics had been selected and broadly defined by the advisory council; workshop participants were directed to realign and develop them further. Each task group was asked to examine its subject in the light of four general questions:

1. What is the urban ecosystem?
2. What do we think we know about it?
3. What do we need to know to improve our understanding of it?
4. How can we apply what we already know?

The groups were encouraged to begin work prior to the workshop; several met for pre-workshop discussions; others corresponded.

The initial phase of the Urban Ecosystem Project culminated in the national workshop held at the University of Texas, Austin, from March 29 to April 4, 1973.

At the workshop, each task group produced a draft chapter for use in this report. During the six days in Austin the task groups attempted to review current knowledge in their subject areas, to prepare their draft statement, and to make recommendations designed to assist those planning, conducting, funding and utilizing urban research. The result of their effort, which forms the substance of this report, reveals the complexity involved in dealing with a completely man-dominated system, the city, from an ecological point of view.

The workshop report is presented in two parts. Part One includes: a brief introduction; the general workshop recommendations, structured to correspond to the arrangement of the full document (Part Two); and a summary view of the urban system. Part One is a readily accessible overview of the workshop's frame of reference, its recommendations, and its conclusions.

Part Two is the workshop report. It is divided into four sections: 1) Goals: values and jurisdictions; 2) Components: population, physical structure, and resource flows; 3) Processes: interventions, indicators, and models; and 4) Case Studies.

Section One is comprised of the reports from the task groups on "Goals for the Urban Ecosystem" and "Institutional Structures". Section Two incorporates the work of the groups studying "Population Processes", "Physical Structure and Function", and "Resources: Water, Energy, Materials". Section Three presents the reports on "Indicators of Urban Ecosystem Function and Health" and "Systems Capacities, Limits and Intersystem Linkages". The final section is the report from the task group on "Case Studies of Cities and Urbanized Regions".

In some instances, material from particular task group reports has been deleted, rearranged or moved to another section. Deletions have been made to avoid unnecessary repetition of material discussed more appropriately in another section. The "Goals" report opens and closes Section One, with the "Institutional Structure" report intervening. The material on populations from "Goals" has been incorporated into the discussion of populations in Section Two. The remarks on federal legislation as it effects urban systems originally part of the "Institutions" report is found in Section Three. The consideration of interventions in the urban system from both "Physical Structure" and "Resources" now appears at the opening of Section Three.

Some overlap in the reports has been retained for two reasons: 1) to show the development of group reasoning and opinion, and 2) to provide a general background to each unit for those readers whose interests lie in one specific section. Hopefully, the overlap is useful. Divergent views, and related evidence from within or between task groups has also been retained to some extent, since they illustrate some of the weaknesses in our present understanding of the urban system.

In combining work by the several task groups into a comprehensive whole, care has been taken not to alter the intent of the workshop task group reports. However, most participants have not had the opportunity to examine the final edited report, nor have they had the opportunity to read the complete text; therefore, it should not be assumed that every participant endorses all the recommendations presented nor that he is in agreement with the conclusions that evolved from a group in which he did not participate.

The national workshop and this report were not intended to be the final step in the development of the Urban Ecosystems Project of TIE. They were to serve as tools available for use by member institutions of TIE to develop urban research programs and to field cooperative teams for urban studies. Several recommendations of the workshop were ready for immediate implementation; these have been provided to specialists and planners in pertinent areas.

The work of the Advisory Council, the Workshop and preparation of this report was supported by National Science Foundation Grant G1-34997. Generous support was also provided by many institutions and contributions of time and effort by unnumbered individuals.

In a project as diverse and complex as this one has been, management of communications and organizational detail became extremely important. TIE and the project director are indebted to Stephen Reeve, project manager, and to Barbara Klopatek, project secretary, for their sustained effort and enthusiasm in carrying out these tasks. Linda Weimer served in the crucial capacity of editor of the draft workshop report; she was assisted by Patricia Mallonee. The project was aided in many ways by Douglas and Trisha McConnell and others of the TIE staff and by Perry Aiken and Linda Thomas.

TIE is indebted to the professional organizations which assisted by co-sponsoring the advisory council and the workshop, and to their representatives who were generous in providing assistance and expertise:

American Institute of Architects,
American Institute of Planners,
American Society of Civil Engineers,
American Society of Landscape Architects.

Several organizations and institutions served as considerate hosts for advisory council meetings:

Johnson Foundation,
University of Wisconsin-Madison,
University of Wisconsin-Milwaukee,
University of Washington.

The cooperation and assistance of U.S. and

Canadian universities and of United Nations representatives were most helpful. The results of the workshop should have value for the 1976 U.N. Conference on Human Settlements to be held in Vancouver, B.C.

Thanks are due to the Canadian Ministry of State for Urban Affairs, the Biology Department of the University of Rochester, the Botany Department of the University of Wisconsin-Milwaukee, and the Biology Department and School of Architecture at the University of Texas, Austin, for their considerable assistance.

The criticisms and suggestions of the T.I.E. review committee have been most helpful. The editors wish to thank John Neuhold, chairman, and the members of the review committee, John Hultquist, Michael H. Levin, Jaro Mayda, Frederick E. Smith, and Eugene P. Wenninger, for their insights and encouragement. We wish also to thank Daniel Carson, Conrad Istock, David Jameson, and John Ong for their enlightening comments on the text.

Special thanks are owed Josephine Doherty of the National Science Foundation for her thoughtful assistance and encouragement in all phases of the project.

It would be an impossible task to acknowledge the many and diverse contributions by participants, members of the advisory council, task group leaders and others, all of whom clearly attached high importance to this examination of the urban ecosystem concept. This volume is the result of complex cooperative effort. It is our hope that it meets some of the needs and objectives envisioned by those who have so generously encouraged and assisted the project.

Forest W. Stearns

Urban Ecosystems Workshop

PARTICIPANTS

Goals for the Urban Ecosystem

Arthur Boughey
Daniel Carson
David Erickson
John Hultquist
Warren A. Johnson
Thomas MacCalla
Clifford H. Patrick
Erik A. Svenson
Linda Alvin Thomas
Robert Woerner

Institutional Structures

Lloyd Axworthy
Hugh Calkins
Alan D. Chambers
Robert T. Coles
Ralph W. Conant
Linda Harris
Edgar Horwood
Chandler H. Stevens
Robert Warren
Robert Williams
Harold F. Wise

Physical Structure and Function

Adolf Ciborowski
William Garrison

James E. Hackett
Kingsley E. Haynes
Malcolm K. Hughes
Norman Hummon
Martin Redding
Chandler Smith
James A. Veltman
James Wuenscher
Arthur J. Zeizel

Resources: Water—Energy—Materials

George Cox
Craig Davis
Edward Deevey, Jr.
William Hart
Ingrid Lundholm
Daniel Phelps
D. H. Waller

Population Processes

Alan M. Beck
Alexander Bigler
David E. Davis
Joan Draper
David Jameson
Charles Lewis
Reynell Parkins
James B. Pick
J. Barkley Rosser, Jr.
Norman Woldow

Indicators of Urban Ecosystem Function and Health

Robert Anderson, Jr.
Sinclair Black
Richard Bradley
William Cooper
Ralph d'Arge
Robert Murray
Theodore Roefs

Robert Stuart
Henry Vaux
Anthony Wilden

Case Studies of Cities and Urbanized Regions

Aryeh Cooperstock
Clyde W. Craven
Robert Dorney
Amos Eddy
Robert Einsweiler
Peter Jacobs
Royce LaNier
C. Ray Smith

Systems Capacities, Limits and Intersystem Linkages

L. J. Bledsoe
Kent W. Bridges
Tim Campbell
Richard Duncan
Charles Ervin
Peter House
W. Cecil Steward
James Zucchetto

Task Group Leaders

L. J. Bledsoe (Systems)
Daniel Carson (Goals)
Ralph d'Arge (Indicators)
Robert Dorney (Case Studies)
William Garrison (Physical Structure and Function)
Edgar Horwood (Institutions)
David Jameson (Population)
D. H. Waller (Resources)

Workshop Steering Group

Josephine Doherty
Patrick Horsbrugh

Conrad Istock
Bassett Maguire, Jr.
Robert Mather
M. B. McPherson
William Rees
Stephen Reeve
Forest Stearns

Advisory Council to the Urban Ecosystems Project

William R. Burch, Jr.
Ralph d'Arge
Robert Dorney
Lawrence Heideman
Patrick Horsbrugh
Edgar Horwood
Conrad Istock
Richard Lee
William R. McGrath
Brian Mar
Stuart Millendorf

Representatives of Co-operating Organizations

Don Conway (American Institute of Architects)
M. B. McPherson (American Society of Civil Engineers)
Alan Rabinowitz (American Institute of Planners)
Paul Sedway (American Institute of Planners)
Robert L. Woerner (American Society of Landscape Architects)

University of Texas Task Group Assistants

Susanne Abright
Martha Bennett
Ellen Boozer
Graham Hunter
Jim Munch
Carol Nathan
Ruth Parshall
Mary Lee Plumb

RANN/National Science Foundation

Josephine Doherty

T.I.E. Project Director

Forest Stearns

T.I.E. Project Manager

Stephen Reeve

Part One

Introduction

The Urban Ecosystem: A Holistic Approach is the result of an intensive interdisciplinary examination of the city. This two-part document is the work of approximately 90 specialists from various natural and social sciences who gathered to discuss the city and its problems in a new and, hopefully, useful way, i.e. as an ecosystem. In the past, interdisciplinary work in urban studies has been limited and the "ecosystem" principles which may underlie the structure and functioning of the urban system have received only incidental attention.

The initial and primary assumption of the workshop organizers and participants is that to view the city as one would view a natural ecosystem may provide useful and valuable insights into the nature and problems of American cities, and that, further, such an approach may lead to solutions to urban problems.

The Urban Ecosystem: A Holistic Approach provides little "new" data on the city; rather, it attempts to correlate and integrate existing knowledge. The book, in providing a framework for future urban studies, attempts to highlight and clarify the principles underlying urban system function. As a breakthrough effort to create a framework appropriate to comprehensive urban study, the book is an early step towards definition of a new multi-disciplinary field of study. Accordingly, it may well contain some internal inconsistencies and some sketchy or otherwise inadequate discussion. Such deficiencies are perhaps inherent in the formulation of any new interdisciplinary methodology. We believe that the framework offered here will serve as the basis for and a spur to further interdisciplinary urban research. The framework, if it proves valuable, can be amplified and refined on the basis of new data and new understanding of urban system functioning.

In addition to providing a fresh view of the city, this document is intended to provide guidance in the selection of urban research priorities and in the allocation of research funds. Accordingly, we offer several general recommendations (in Part One) and numerous specific proposals (in Part Two). We urge that these recommendations be carefully considered by the appropriate policy-making and funding agencies.

Our report is divided into two parts. Part One presents the major recommendations resulting from the workshop and an introduction to the ecosystem view of the city. The recommendations here are grouped and structured to correspond to the arrangement of the full discussion in Part Two. Those charged with selecting research priorities and with funding research projects can look to the major recommendations to find the broad areas that workshop participants believe require urgent attention. Each policy, research, or implementation recommendation in Part One is preceded by a brief discussion intended to place it in context. Fuller understanding of the major recommendations can be gained by a thorough reading of the appropriate detailed discussion in Part Two.

The concluding section of Part One, while not a precise summary of the fuller discussion in Part Two, serves to orient the reader to the ecosystem view of the city and to the principles underpinning the fuller treatment in Part Two. This section briefly touches on many aspects of the city which receive more detailed consideration in the second part of the report.

Part One, then, is intended to make the major recommendations and the general orientation of the full document easily accessible to governmental problem-assessment agencies, to those evaluating research priorities, to budget planners, to congressional staff members, to urban researchers and planners, as well as to the interested lay audience including public officials and citizens' groups. The reader, lay or professional, is invited to use Part One as a guide to the fuller discussion of the urban system that appears in Part Two.

Part Two of *Urban Ecosystems: A Holistic Approach* is divided into four sections 1) Goals: values and jurisdictions; 2) Components: population, physical structure, and resource flows; 3) Processes: interventions, indicators and models; and 4) Case Studies. A number of recommendations follow each section; these may be of special interest to urban research or design professionals.

The specific proposals arise from the detailed discussion; they recommend changes in existing structures or policies, outline new approaches to and uses for urban system information, or indicate particular areas which require further research. The proposals are grouped, in each section, as: policy recommendations; recommendations for action; and research recommendations. Although some of these recommendations were offered by one or more task groups for inclusion as recommendations in this document, most of the specific proposals were drawn from the text by the editors and fashioned into recommendations.

A list of references is found at the end of each section of Part Two.

Section One of Part Two treats two main

areas of particular importance in understanding urban systems: values and institutions. The urban system has a twofold goal: to sustain human life and to enhance it. Our values, reflected in our behavior, and our institutions greatly influence life in the city; they affect both urban system functioning and the formulation and implementation of urban system policies and programs. Understanding the nature of human values and human institutions will assist us in formulating ecologically sound urban system management policies and behavior, and will, hopefully, help sustain and enhance human life.

Section Two deals with the components of the urban system: the urban population, urban structure and function, and urban resource flows. Urban population is a particularly significant component of the urban system; the city exists to serve its residents, yet the density and distribution of that population greatly influence the city's capacity to provide essential services. Non-human urban populations, often neglected in urban studies, are given detailed consideration here. Structural and functional constraints often limit the city's capacity to provide services to its residents; understanding urban structure and function, and their interrelatedness, should help improve urban design and planning and may ultimately serve to enhance urban life. Likewise, understanding of essential resource and residuals flows in the city should contribute significantly to attempts to improve the urban condition.

Section Three details possible interventions into urban structure, functioning, and resource flows, discusses the indicators necessary for assessment of urban system health, and proposes a systems model helpful in understanding the actual urban system and its processes. Intervening into urban processes requires adequate assessment of system health and adequate techniques for understanding the way the system functions; a set of clear and useful indicators and a comprehensive model of the urban system can be immeasurably valuable in changing and improving life in the city.

Section Four discusses six approaches which have been used to examine urban systems. These case studies range from a brief overall survey to extensive systems studies.

We are well aware that the ecosystem approach is not the only approach to urban systems. We trust, however, that the approach will aid understanding of the city and contribute useful information and, possibly, solutions for urban problems. Certainly, this work is not meant to be a definitive statement on the urban system; rather, it is a beginning.

Major Recommendations

I. Section I: Goals: Values and Jurisdictions

1. Behavioral changes in individuals and institutions may aid achievement of societal goals. Social change can be brought about by legitimization, persuasion, or coercion. Actions intended to change urban society should result from general societal consensus; these actions must be implemented early enough to avoid inherent system delays and need to be uniform and objective enough in application to accomplish desired changes.

The dangers involved in instituting social change include: 1) the goal sought may prove short-lived or undesirable; 2) the change may produce unexpected system responses, perhaps contrary to the desired goal; 3) the action may not have sufficient force to carry it to fruition; thus, the action and its means for implementation must be adequately funded and supported by the urban community.

Policy Recommendation. We recommend that high priority be given the development of specific social means for implementing the goals of the urban society. Changes in attitude, behavior, and life style should be encouraged through education.

2. Environmental research is often confined to the physical aspects of the system in which

man functions. Such limited research minimizes the significance of man interacting with an environment which includes other men, value conflicts, social inequities, alienation, and stress.

Research planning, policy-making or institutional arrangements for human settlements should consider both the psychological and biological needs of man. The nature of values is poorly understood and methods of incorporating value systems into community planning are poorly developed.

Research Recommendation. We recommend research on the nature of values, with particular attention to: 1) their distributions and strengths in human populations; 2) the causes, rates, and directions of value development and change; 3) value orientations and conflicts; 4) the roles values play in choice; and 5) the effects of satisfaction and/or frustration, stimulation and stress on value systems.

We further recommend research on and implementation of methods to more effectively incorporate consideration of fundamental value systems (current and potential) of the population in community-based planning.

3. Experience in the United States indicates that well-conceived public projects are often not accepted; rejection by the public largely results from insufficient information and understanding. The projects lack the necessary effective public commitment. Investigation of the processes of community involvement should help develop procedures which produce more acceptable recommendations and which allow those affected a voice.

Research Recommendation. We recommend research and testing to improve processes for the involvement of the general public and citizen groups in the formulation of public policy decisions. Such research will require continuous, cooperative participation from behaviorists, planners, environmental scientists, politicians, urban administrators and citizens. Suggested techniques must actually be tested in true-to-life situations; every effort must be made to prevent abortive trials resulting from co-option of a research program or field trial by special interest groups or special purpose governmental agencies.

4. Various social indicators suggest that ecological problems are not perceived similarly by the poor as by those whose basic wants are satisfied. To design functional institutions and insure their use, we must know whether the various segments of our population have significantly differing perceptions of the ecological matrix and whether these segments exist at different levels of ecological imbalance.

Research Recommendation. We recommend research into the perception of ecological problems by various segments of our society.

5. Each natural region has its own particular set of opportunities and constraints for development. Such inherent characteristics must be recognized in regulatory and policy measures guiding land use and population distribution. For example, natural environments incompatible with urban development should be identified and protected from over-development or destructive use.

Policy Recommendation. We recommend that state and national urban growth policies be developed promptly and that they recognize and reflect ecological sensitivities, capacities, and regional characteristics.

6. Examining performance criteria in land use policy will facilitate inclusion of the concepts of carrying capacity, threshold and lag effects, and amenity resources. Today, we use blanket environmental standards and zoning to control land use. These should gradually be replaced by performance criteria which can include socio-economic variables, and which take into consideration variation between landscape units and replace residual control with a flexible policy, thus conserving energy and mineral resources.

Policy Recommendation. We recommend that performance criteria be gradually substituted for land-use standards and zoning. Research is required to establish and test such criteria.

7. The current rate of economic growth may require curbing by deliberate action; immediate attention should be given to the short and long-range consequences of such action. The consequences of our attempts to maintain present growth patterns have likewise been inadequately considered. Either deliberate change or maintenance of present growth patterns will have economic, environmental and social effects.

Economically, growth maintenance or pattern change will affect future resource availability and standards of living. Environmentally, either will affect productivity, stability, and human health. Socially, either will affect the ability of individuals with various racial and ethnic backgrounds to achieve full participation in society.

Research Recommendation. We recommend comprehensive analyses of the economic, environmental and societal consequences of alternative rates of economic growth within urban systems of the United States.

8. A number of desirable changes are occurring in the urban ecosystem, largely as grass roots political responses to a changing social and natural environment, to the increasing scarcity of energy, and to the uncontrolled growth of cities, etc. Often, efforts to cope with urban problems create controversy and opposition to the environmental view of the urban system and the best interests of the public. Facilitation of the ability of "change agents" to promote achievable objectives and to reduce contradictions may allow necessary or desirable changes to occur more quickly.

"Change agents" include these, among others: groups or individuals developing small scale technology; watchdog groups—to see that industry obeys environmental laws; groups trying to preserve neighborhoods, historic locations and parts of town or other valuable environmental amenities.

Implementation Recommendation. We recommend that public funds be allocated to various "agents of change" in our society to enable them to establish a sounder basis for their efforts, to reduce socially disruptive conflicts in their programs, and to facilitate their ability to promote political change.

II. Section II: Components: Population, Physical Structure, and Resources

9. Poverty, unemployment, crime, drug abuse, alienation, racism, mental disturbance and chronic mobility are all indications of population pressures under conditions of deteriorating social opportunity.

Behavior is a major adaptation to environment. Behavioral response determines our quality of life by mediating human interactions, health and perception of well-being. Furthermore, behavior is the means of accomplishing changes in the relationship between people and their environment.

Recorded data on the various parameters of the urban social crisis are already available. They need assembly, analysis, collation and study by multidisciplinary teams working within a short time span of 2-3 years.

Results would provide immediate feedback over a wide social field, including population growth and movement, education and vocation opportunity, dwelling and urban design, job and resource allocation, and recreational and health facilities.

Research Recommendation. We recommend increased attention to the behavioral problems of urban areas, including research on: behavior in relation to space and crowding, economic status, and available opportunities; the determinants of aggressive actions; behavioral aspects of environmental parameters and health. These should be examined in the context of physical, biological and psychological resource limits within urban ecosystems. Population density as it affects demographic and social interactions, including fertility, mobility, stress and stimulation deserves special attention.

10. Changes in fertility patterns influence a population's resource consumption. The birth rate also affects social services, economic structures, transportation networks, and cultural patterns. Prediction of fertility patterns and knowledge of their consequences will improve our capacity to plan.

Research Recommendation. We recommend increased research to determine the effects of fertility patterns on age structure, and their resultant impact on resource use and distribution and on planning and design of urban environments.

11. Planning decisions made for the local population or a local area are easily disrupted by the in-migration of unexpected numbers of people or by sharp alterations in the flow of capital and goods. At present, we cannot adequately assess the causes or the size of these migrations or the socio-demographic profiles of the migrants. Large population shifts can be particularly disruptive. The primary constraints on large shifts of population, wealth, or goods are economic. In-migration of people has led to the decline of local environments and the local quality of life when local populations must rapidly begin to support substantial numbers of migrants. We place a high value on each individual's freedom to move, but this freedom now plays against declining local living conditions. Communities feel the need to better control migration, but they lack methods of control consistent with accepted standards of American liberty.

Research Recommendation. We recommend expanded research on the causes of human migration into, within, and among United States' cities. The causes of large economic shifts and the predictable consequences of such shifts must be examined. Further, research is needed to investigate ways in which the detrimental effects of human migration and economic shifts can be reduced.

12. Elements of the present urban system are often segregated on the basis of socio-economic status (based primarily on income). Such separations leave many urban residents in slum conditions, may encourage crime, and may cause mental stress. Many residents are denied opportunities available to the more affluent.

Further, interaction and feedbacks among and within the formal and informal sociocultural, race-ethnic and religious institutions have led to the subordination of long-run social and humanistic values to short-run economic values.

Research Recommendation. We recommend investigation of the degree to which present income distribution may result in: 1) denial to racial and ethnic minorities of opportunities for decent housing, health care, education and political and legal representation; and 2) large migration streams from areas of less opportunity to areas of greater opportunity.

13. Race and racism create a variety of social and environmental problems in the cities of the United States. These problems reduce the quality of life and the opportunities of a significant number of urban citizens. The United States has the resources to alleviate racism, but has not shown the determined and effective commitment to do so. Specific proposals to eliminate the effects of racism are urgently needed.
Research Recommendation. We recommend greatly extended research on the ways racism continues to influence urban life. We further recommend research on ways to alleviate the social and economic consequences of racism.

14. Scattered evidence suggests that plants and animals may serve numerous functions for the physical and psychic well-being of man. The effects may range from shading to SO_2 absorption, to the stimulation of childhood mental development and to the easing of stress in humans of all ages. Other evidence, likewise limited, has demonstrated the dangers to health and safety caused by plants and animals. Plants, animals and other natural objects may influence human behavior in ways yet unknown.

Research Recommendation. We recommend that increased attention be focused on the relationships between human populations and populations of plants and animals in urban areas. Such research should involve consideration of both the positive and negative aspects of various aesthetic, psychological and physical interactions, as well as those relationships which affect human health.

15. Self-maintaining natural systems provide services to man without human expenditure of energy or resources. For example, flood plains or cypress swamps may provide flood control and aquifer recharge; marshes supply nutrient to estuaries; mangroves may prevent erosion; woodland provides protection from the sun and may reduce air-conditioning costs; huge quantities of air and waterborne wastes are processed by forests, marshes, lakes and streams. As yet, our knowledge of these natural processes is fragmentary.

We do not know the extent to which natural areas actually perform such functions; the relative efficiencies of alternative uses of open space of different sizes and with differing cover types and arrangements are not yet known. Data on these functions would allow economic comparison of natural system costs with the costs incurred in using man-made structures to perform the same functions.

Information on natural system functions adequate to prevent system damage could result in new techniques for building integrated communities of man-made and natural systems characterized by diversity, long-term stability, and aesthetic attractiveness. Approaches to environmental planning utilizing such knowledge have already begun in areas of Florida as part of accepted rational growth policies.

Research Recommendation. We recommend research to identify and quantify the services performed for man by urban parks, wetlands, undeveloped areas and other natural systems. The ways these unmanaged communities can—in terms of size, vegetation and spatial arrangement—provide nonsubsidized work for man demand examination.

16. Every system, whether completely "natural" or largely human-oriented, is subject to outside disturbance and to potential failure of some or all of its components and their functions. Independent ecosystems develop structural diversity and functional redundancy as they mature. Whether or not and in what form increased diversity contributes to the stability or resiliency of a system is a controversial issue currently undergoing scrutiny in ecological research. The possible mediating effects of diversity in urban systems are not understood.

Research Recommendation. We recommend research to determine the positive and negative implications of redundancy and diversity of form, function, and population composition in both the administrative and physical structures of urban regions.

17. Urban ecosystems consume enormous amounts of energy. Various combinations of city configuration, population density, differentiation and unit connectivity presumably have differing energy needs. The relative energy efficiencies of alternative combinations are only grossly known. Energy conservation is already a matter of urgent concern, but we continue to permit the growth of cities without regard for their energy efficiency.

Research Recommendation. We recommend detailed examination of the structural design of cities to determine their efficiency in energy utilization. Such studies would be aimed toward conservation of energy in both new and renovated urban systems.

18. The rates of resource depletion and the accumulation of biologically harmful residuals discharged from the urban system are rapidly increasing. Problems of production, consumption, and residuals discharges have usually been treated separately, leading to the optimization of one segment at the expense of the system as a whole. Existing knowledge of environmentally sound production and discharge processes has not been fully implemented; further, present means for stimulating adoption of such processes appear ineffective.

Research Recommendation. We recommend high priority for research on the implementation of alternative techniques for reducing resource consumption and residual output in urban systems.

III. Section III: Processes: Interventions, Indicators, and Models

19. The present market system is based on private, short-run valuations of resources rather than any long-run social valuation. The result is under-pricing and inefficient transformation of scarce resources, leading in turn to pollution. Social valuation, "internalizing the externalities", would help reduce the destructive transformation of scarce resources and the resulting pollution, both of which threaten long-term human survival. Alteration of the market system requires understanding of

the interaction and feedbacks among and within the economic, political and legal institutions as they have evolved and influenced the urban ecosystem.

Policy Recommendation. We recommend investigation of the degree to which: 1) present income and property taxes could be changed to reduce rates of resource use; these rates have resulted in depletion, pollution, over-capitalization, abandonment of the central city, and movement of industry and population from areas of high taxation to areas of low taxation; and 2) economic solutions to various pollution problems actually conflict with political and legal realities.

20. Residential development in a dispersed pattern appears to have adverse effects on the natural environment, on the ability of the residents to establish a sense of community, and on the ability of the lower-income groups to obtain homes. Are these effects actually increasing? Are there other positive or negative features of residential dispersion? We need information on these phenomena; if dispersion is generally detrimental, efforts to reduce dispersion require strong support.

Research Recommendation. We recommend broader and more intensive investigation on the ways residential dispersion (especially as found in newer residential developments) effects the use of land and resources, transportation, pollution, parks, open space and human behavior.

21. Funds spent on urban ecosystem research and design and on huge data banks alone will yield little relevant information on solutions to urban system problems. Applied research on urban systems lacks a unified basis of principles and is now inefficient. Immediate investment in indicators research, emphasizing long-term predictability, may prove highly productive.

Policy Recommendation. We recommend that research to monitor urban ecosystems be concentrated on the identification of a few aggregate indicators for environmental quality and health. Such indicators should offer long-term predictions of environmental security and stability.

22. The merging of natural and social sciences is overdue and would benefit both disciplines. To accomplish such a merger, a translation matrix is needed to aid and stimulate researchers and to provide clear warnings against blind mapping of one science onto another.

In specific cases, generalizing a concept or principle may permit comparisons between natural and urban ecosystems. Considerable care must be taken to distinguish between analogy and identity.

Research Recommendation. We recommend an exhaustive comparison of urban and natural ecosystems. Such a comparison will require a listing of laws, principles, concepts, phenomena (understood or not), definitions, etc., of both natural ecosystems and urban ecosystems. Similarities and differences between two systems should be itemized in detail. Such comparison should aid the development of a translation matrix needed for interdisciplinary studies.

23. We urgently need alternatives to present technologies supporting urban functions, i.e., shelter, waste disposal, planning, and

transportation, etc. New technology must be stimulated, designed, tested, developed and finally analyzed for its long range effects.

Much research is already underway on these topics, but the effort is small compared to the need. There is little or no work with the breadth of ecosystem analyses; systems alternatives rarely receive priorities in research.

Research Recommendation. We recommend support of extensive multi-disciplinary work toward new, low energy technologies to replace or supplement those presently used in the urban system. Work should not stop with identification and design of such technology but must be carried through to test and analyze these technologies from the ecosystem view.

24. The complexity and size of large scale, computer-based ecosystem models, especially those including urban processes, frequently preclude familiarity with the hypothesis mechanisms, data reliability, implementation and results of these models. This complexity creates significant problems in publication and dissemination of the results of such modeling efforts. Assessing the usefulness, accuracy, and completeness of such models requires their communication and reproducibility by others. Review mechanisms must be developed by providing explicit procedures for dissemination and duplication when such large modeling efforts are undertaken, especially in conjunction with urban studies.

Policy Recommendation. We recommend that projects proposing to model all or significant aspects of an urban ecosystem include mechanisms permitting independent assessment, replication and evaluation.

25. The environmental information useful for political decisions varies depending on the scale of the decision. Decisions may range from federal funding for a new town of 250,000 or larger metropolis to city funding for widening a street. The need to provide accurate technical information on environmental function, cost, carrying capacity, aesthetics, etc., at different administrative levels requires different levels of scientific sophistication and different techniques. Systems analysis appears potentially useful at the national and state levels while overlay mapping of natural features and amenities of proven utility should be useful at a finer scale for county and state level.

The destructive potential of systems analysis, if used for private gain or if misused by government, is great and requires independent "audit" by an agency of government. Only in this way can we provide for adequate design and application of advanced information and analytical technologies as they come into general use.

Implementation Recommendation. We recommend the establishment of a hierarchy of coordinated environmental information and analysis services from federal to local and city levels, for different kinds of environmental problems. At the same time, a more general auditing agency is essential to ensure that sophisticated data banks and environmental analysis systems are not exploited by special interests to the detriment of the public interest.

IV. Section IV: Case Studies

26. City and county governments generally depend upon natural resource information from local, state (provincial) and federal bodies; these bodies, in many cases, do not have the appropriate information, (i.e., scale of mapping, suitable sensitivity of analysis or

adequate sampling network) for reaching wise decisions affecting land use. A small local team of environmental scientists and designers could conduct surveys, reinterpret available information to proper scale, interact with, and provide information to citizen groups, and government agencies on issues of environmental quality. This team would reside on "neutral turf", adequately buffered from city hall to be credible to citizens groups, yet close enough to politicians to be useful as a technical resource. Such a team should be linked to environmental agencies at the state, regional and federal levels which have systems analysis capability. The environmental team concept has certain similarities to the Agricultural Extension Service which opened channels of communication among farmers, government and universities.

Implementation Recommendation. We recommend that teams of environmental extension personnel of varying disciplines be established in urban and urbanizing areas; the teams should be supplemented at the state level by interdisciplinary terms using computer and systems analysis; the procedures and results of the teams' studies should be fully accessible to governmental bureaus and to citizens. These teams, providing an extension function, could serve as a potentially crucial link in environmental design and management, as well as in public communication.

Towards The Urban Ecosystem[1]

[1] Prepared by Conrad Istock, William Rees, and Forest Stearns.

Technically trained persons have their own characteristic biases, and if their view of the city is different from that of the commuter on the Long Island Railroad it is not necessarily more realistic. Moreover, as the technician comes to play a more important part in policy-making, he is bound to come more and more under the discipline of large organizations, especially foundations and government agencies, whose maintenance and enhancement depends in some way upon the elaboration of an alarmist, or at any rate expansionist, public definition of the situation. That young newcomers to the small world of opinion-makers tend to be open to new ideas is not altogether reassuring either, for they may tend to accept new ideas just because they are new. To the pessimist, the prospect is that a new conventional wisdom about the problems of the city, the product of many millions of dollars expenditure on research, cast in the language of systems analysis and the computer, will only compound the existing confusion. The optimist, however, will see reason to believe that facts, rational analysis and deliberation about the nature of the public interest will play a somewhat larger part than hitherto in the formation of both opinion and policy.

E. C. Banfield
The Unheavenly City
1970

We may gain new understanding of the condition of American urban areas if we begin to treat the city as a "system", or as a part of a larger, regional "ecosystem"—a view which emphasizes the fundamental biological nature of man interacting with the complex urban system. Such treatment is the focus of this report.

The "urban ecosystem" approach may not solve all urban problems, but its logical viewpoint is certainly useful to the study of cities, for it reveals their actual complexity and makes clear the urgent need to have urban planning, decision-making, administration and research become more interdisciplinary in nature. Many urban programs have failed because the concepts on which they were built were not sufficiently comprehensive and integrated. The urban ecosystem approach may provide a model comprehensive and integrated enough to meet the needs of our cities.

Some contend we can solve current urban problems merely by more money and wiser use of technology; however, most of the nation's energy and resources already go toward maintaining the urban way of life. By viewing the movements of energy and materials into and out of American cities as general ecosystem processes, we see even more clearly that the relationship between those movements and the quality of urban life is not simply an economic one. The United States, with its higher levels of resource use, higher per capita GNP, higher per capita energy consumption and lower population density than the nations of Western Europe, appears to have the more severe urban problems. Higher levels of resource use and economic activity do not automatically raise the quality of urban life but may instead cause it to decline.

One might speculate that the U.S. industrial machine now dictates its own needs, requiring man to satisfy its ever-increasing appetite for energy, materials and human labor. Actual goods and services produced seem a secondary consideration (Galbraith, 1967), diverting resources and human efforts from activities which could meet human needs and enrich the quality of life in America. One hopes we can still muster the resources necessary to follow a more beneficial and sustainable pattern of development.

If resource supplies and economic growth rates should decline, we may find urban conditions worsening. Although this is a matter of considerable debate (Beckerman, 1971; Ayres and Kneese, 1971; Istock, 1971), the nature of the outcome will more likely be indicated by real world experience than by further speculation. But the debate does highlight the need for more effective methods of anticipating future requirements and of allocating present resources more wisely.

At the minimum, the U.S. population is expected to increase by 60 million people in the next thirty years. At present standards of living, such growth will require the construction of housing equivalent to that in New

York City every five or six years to meet the needs of these new citizens. Many of those in the higher economic levels of our society, presently living outside the most troubled parts of our cities, can adopt for a time a detached and philosophical view of urban decay. However, their stake in the city's future becomes clear and the imminent need for urban rehabilitation apparent when we consider the nation's expected growth and the changes that will ensue.

Our challenge is to meet present and future needs without continuing to destroy our land, water and air. The immensity of projected growth provokes some to wonder whether we can even hope to have habitable cities thirty years from now, and whether we would be better off without cities. But urban problems are not isolated from the human condition, nor from the condition of the entire nation. The future of our cities is inextricably linked to the future of our country. Urban problems thus demand our immediate attention.

CITIES

Cities have long been centers of collective human activity. Ancient city-states were the administrative, military, economic, legal, educational and artistic centers of the geographical areas they ruled and influenced. The first cities probably arose as man moved from small-scale family or tribal agriculture to more productive, stable and predictable agricultural systems which fostered diverse and specialized human occupations far removed from the tilling of the land yet dependent upon and sustained by it. Today, even those involved in the most sophisticated human activities retain this ancient tie to the land, a fact often forgotten. We are completely dependent upon the productivity of the land, and it is this relationship which is perhaps the most basic one existing between the city and its biological setting.

Since the beginning of the Industrial Revolution, cities have undergone the most rapid and profound changes ever experienced in their ten or twelve thousand year history. Man has covered vast areas with buildings and other construction; integrated systems of exchange and services were set in place; layer upon layer of government were added to coordinate, regulate and plan the city's activities. These cities provided the highest levels of nutrition, shelter, longevity of life and opportunity ever known to man. However, with the immense growth a host of urban problems has intensified: overcrowding, crime, unemployment, racial tensions, declining quality of urban life. The question poses itself: what is causing the deterioration of our cities and how can we stop it?

We are a long and complicated way from the city's agrarian beginnings. But the basic biological roots of human society remain, and

we must look there to improve our understanding of America's urban crisis. Though the ecological approach alone cannot render the city entirely comprehensible or provide full solutions to all urban problems, some aspects of the city—human population processes and urban plant and animal interactions, for instance—are best approached ecologically. Likewise, the capacity of the "natural" environment to absorb and process the by-products of urbanized society is primarily an ecological matter. Further, though only in a metaphorical sense, we can compare the flow of energy and materials though the urban system to the biological exchanges of energy and matter in natural ecosystems free from human influence (*Man in the Living Environment,* 1972). When discussing human behavior and the city, our analysis must include values and perceptions, goals, human institutions and the political systems which govern individual, group and institutional behavior.

Clearly, we do not understand the whole city as an operating "system"; necessarily, analysis of the urban system goes well beyond the usual technical meanings of *ecology* or *ecosystem* and beyond the bounds of any particular academic discipline. The ecosystem view is offered here as a useful and necessary approach for fathoming the urban system. It represents a general framework within which the talents and experiences of many disciplines can be brought to the task of penetrating the total urban complex. Some striking conclusions emerge from such a view.

*Individual survival and the survival of urban centers, as goals, would probably receive nearly universal approval in our society. Survival was not a central issue during the development of cities over the last two decades. Today, however, with rising threats to our highly integrated cities and their far-flung life support systems, survival has gained a new and central position in our sets of values and goals.

*Beyond survival, agreement upon social values and goals is difficult to achieve. As a nation, we are far from agreement on the desirable nature of social, institutional and political structures aimed at seeking and establishing common goals. Perhaps many can agree on the goal of enhancing the quality of urban life; but, almost certainly, "enhancement" takes on vastly different meanings for different classes and different racial, ethnic, religious, age and sexual segments of the urban population.

*Historical, physiographical, geological, ecological, climatic and cultural features have made American cities vastly different from one another. Different cities, therefore, may require different solutions to similar problems; case studies of even a few cities reveal these differences and indicate that each city must, to some extent, be treated as a special case when solutions to its problems are sought. Such differences, of course, complicate matters of analysis, policy design and implementation, and make apparent the need for integration of responsibilities for planning and management at local, regional, state and federal levels.

*In recent decades, informal institutions such as community service organizations, environmental action groups, and consumer protection groups, have proliferated. Many of these groups have proven powerful forces, applying information, money, litigation and human energies to certain issues; many have risen to stand cheek to jowl with our formal institutions. Such action-oriented groups indicate both the public's dissatisfaction with the way formal institutions are handling their affairs and a fundamental optimism that things can still be changed for the better.

*Many natural and social processes are linked together to drive or control the flows of energy and matter from their sites of natural occurrence through the city to points where they are either totally dissipated or returned to the environment. These flows of energy and materials must be described quantitatively and planned for, so they can sustain not only modern industrial agriculture and the flow of food to the city, but also water supply, waste treatment, transportation, basic industry and the gamut of social services essential to the city.

*The flows of energy and matter, if they are to support urban life with minimal release of wastes into the environment, cannot be isolated and considered one by one, but must be treated as parts of the total urban system. This approach requires monitoring the present state of such systems and accurate assessment of their future tendencies; it requires also that we use certain indicators to characterize the social and environmental condition of a city and its surrounding area. Social indicators present a particularly troublesome requirement. We have traditionally relied on national economic trends as our closest approximation of such indicators; but in repeatedly failing to reflect many important social costs they reveal their greatest inadequacy. We are better able to devise indices of environmental quality (such as dissolved oxygen levels in water or suspended particulates in air), but these likewise fail to assess the total present or future state of an urban system. Without faithful indicators of the total urban system, adequate planning and regulatory schemes are difficult to devise.

*Many hope the interaction between social and environmental systems will become balanced and self-regulatory over time, behaving much as undisturbed ecosystems; many hope such self-regulatory urban processes will emerge from individual decisions made in buying and voting and thus from economic tendencies and political decisions. However, our economic and political systems are badly matched and poorly responsive to one another. Given such conditions, solutions to urban problems are unlikely to emerge spontaneously from these systems.

Problems of Cities

Although cities still deliver physical essentials—water, food, services, and energy—to residents, they seem to work less efficiently, or not at all, in many other respects. The special problems cities face can be discussed within four general categories:

Socio-demographic trends. Cities accumulate the poor and disadvantaged who find themselves isolated in the central city. Problems of crowding, housing, health care delivery and antisocial behavior, for example,

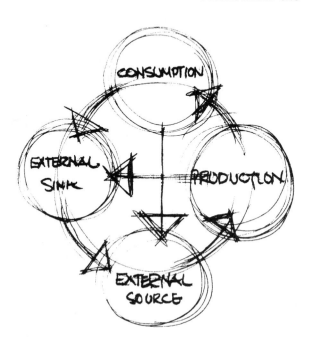

are intensified in the urban core and are determined or influenced by population processes. Further, non-human animals and various plant species, important components of urban populations, are frequently over-looked.

Limits of Urban Growth. The input of resources and energy in our cities is far better managed than the output, largely heat and material wastes. Many cities are now approaching their limits in accommodating their populations—approaching their "carrying capacities". These limitations, as well as impending resource shortages, require not only that inputs be better managed, but that outputs are minimized or better controlled as well.

Wastes. The problem of wastes in an urban system, once a problem only of what to do with them, has become the problem of what they are doing to us. Wastes are having serious visual impact on urban landscapes, and—more seriously—are leading to the accumulation of long-lived, toxic pollutants which can adversely affect all organisms in the urban system and beyond.

Movement of People and Materials. Apart from geographic features, transportation systems and technology give urban centers their form; we can reasonably assume this will continue to be the case. Planning of transportation systems with more conscious awareness of the impact they will have, therefore, becomes an important facet of planning for the total urban system.

Where man might intervene to solve these four sets of problems and what effect his intervention is likely to have become the particular considerations of the following detailed discussion.

Socio-demographic trends

Population growth and distribution are questions which cannot be separated from the most immediate symptoms of urban decay. Total population growth generally increases pressures on metropolitan areas; heterogeneous population distribution, with the poor and disadvantaged located in central areas of the city where problems are intensified, is evidenced in almost every facet of urban life, from the social and political to the economic and ecological. Decisions by subgroups other than those dwelling in the central city are generally responsible for such population distributions. Urban dwellers are left little freedom of choice, and even the well-to-do find themselves restricted in many ways.

Better knowledge of the effects of crowding among urban populations, now poorly understood, would be useful in countering the continuing trend toward concentration of low income groups within the central city. Presently, impoverished parents in the central city, who can observe their counterparts with large families receiving public support, have little direct incentive to restrain their own family's growth, though in the long run it may be to their advantage to do so. Peer behavior and our welfare system—providing support according to family size—tend to encourage the growth of the impoverished class, furthering difficulties. The alternative—providing the poor with the means (jobs and income) to support themselves—might well cost the rest of society less than the amount now spent to maintain welfare programs, insurance coverage and security forces. Such an alternative should be explored.

Urban life tends to place a sensory overload on residents. A variety of avoidance techniques and the screening out of all personal contacts except those most rewarding or least dangerous to the individual are used to counter such overload (Milgram, 1970). Designs aimed at decreasing contact overload while preserving the richness of urban life are a prime and difficult goal. Though urban dwellers have devised ways to cope with the stresses of city living, they have not adapted at a rate equal to the rate of urbanization. Indeed, it is doubtful man could ever completely adapt. Keys to understanding and measuring the effects of city life on human behavior are urgently needed.

The physical and psychological demands of urban living are closely related to the apparent willingness of many residents to passively accept urban conditions. The olfactory, visual and physical assaults of the city overwhelm man's sense of and need for a wholesome environment; acquiescent "adjustment" is actually maladaptive; indeed, sensory overload may cause disinterested and sometimes hostile reactions to campaigns aimed at improving conditions. Further, health problems of city dwellers are greatly exacerbated by urban pollution and population density.

The age distribution of urban population also causes difficulties. An unbalanced age distribution, heavily biased to young and old age groups, exists in many cities. Still, urban managers and planners have largely ignored age structure, except perhaps in their planning for school buildings and playgrounds. Private interests have followed the market, building urban and suburban housing for those in young adult and middle-age groups. Lags in the market system, however, mean that housing needs for the younger groups are barely met and that the needs of the older age group are hardly considered. In many cities, the numbers of older and less productive citizens are growing substantially, as are the requirements for their support and housing. Adequate urban design and increased employment opportunities could greatly alleviate these conditions.

Migration is perhaps the single, most important determinant of urban population patterns. The massive rural-urban migration characteristic of all developing countries, as well as current intercity and intracity migration in the United States, is a manifestation of the search by individuals for greater opportunity and more satisfying surroundings. Study of such aggregate movements reveals patterns of wealth, economic growth and opportunity not made nearly so obvious by other methods. Better knowledge of the causes, scope and age-distribution patterns of migration will aid development of population redistribution policies and successful management of developing or expanding cities.

Plants, animals and insects, which abound in American cities, should not be overlooked in consideration of urban populations. Some such organisms are "unfriendly" pests or carriers of disease; others are plants and animals which provide urban residents some vestige of nature, and with which many people feel a vague yet vital kinship. Urban designers and administrators, however, have tended to concentrate on human-oriented aspects of the urban environment, neglecting the non-human species.

In general, our knowledge of pest control and epidemiology in American cities is sparse. We exercise almost no steady control over rodent and insect pests, nor over a wide variety of debilitating bacterial and viral pathogens. Public health conditions could be improved if intensified studies were directed toward finding ways to regulate population growth or to interrupt the transmission pathways of organisms inimical to human health.

Our understanding of the benefits of non-human urban species should also be extended. Increasing the number of urban parks and green areas and the use of horticultural and ecological knowledge in their design would increase the beauty and the physical stability of American cities, and perhaps their social stability as well. Plant communities more nearly self-maintaining than the present manicured areas of regularly mown grass could be

used. At present we poorly understand the advantageous use of either native regional species or natural growth forms.

Limits of Urban Growth

Most cities can presently supply their residents with the essentials—water, food and energy. Many, however, are approaching their

limits, providing these resources at accelerating cost. The carrying capacities of some cities have occasionally been exceeded. Both energy and water shortages have been recorded in New York; adequate supplies of water to cities are a continuing problem in California.

Food supplies to cities have sometimes failed in other parts of the world, with massive starvation the result. In this country the flows of certain foods—milk, for instance—are frequently limited. Under certain conditions of economic and social stress, such as strikes or riots, food supplies in some cities have been cut off almost completely.

Likewise, urban reliance on energy is now so great that if delivery is interrupted even briefly, the total system is perturbed. Such instances are bound to recur more frequently, as growing populations, increasing energy demand per capita, and urban maintenance requirements further tax available supplies, or as accidents and breakdowns disrupt the distribution system.

Technological innovations may extend the carrying capacity of cities, but these signs of approaching limits suggest that we ought to examine closely the present trends in urban-rural population balance, and perhaps encourage a shift in the direction of migration. We must recognize, however, that any alteration of population distribution may create adverse environments in those areas now urban-supporting, and that such developments may exacerbate rather than resolve present urban difficulties.

Wastes

Wastes—solid, liquid and gaseous—are produced in huge quantities by the city's dense concentrations of people and industry; their disposal is clearly one of our major urban problems. In small quantities, most could be readily assimilated by natural supporting systems—the water, soils, and vegetation of nearby forests, rivers, lakes or grasslands. Concentrations of waste, however, swamp the processing capacities of adjacent systems, causing them to function inefficiently or not at all.

Reducing the rate at which resources flow through the urban system is one way to reduce wastes. We may also reduce environmental effects through product or process change, though such change will not necessarily alter the total amount or kinds of residuals produced. We may choose to treat or store waste materials. Whatever the solution, however, tradeoffs are involved: one resource played against another; one residual reduced at the cost of the increase of another. Almost any attempt to remove or recycle wastes involves environmental changes and economic adjustments; each intervention into the flow system, therefore, requires careful examination.

Intervention often has serious effects on natural landscapes already supporting human populations. Air pollution standards have ended the incineration of urban solid wastes; as a result, the development of sanitary landfills has been brought to a fine art, though acquisition of sites is increasingly expensive and difficult. In many areas, the reduction of urban air pollution through the filling of lowland sites—of little immediately obvious value for building or agriculture—has meant the destruction of wetlands and their less quantifiable usefulness as wildlife habitat, as nurseries for many important marine food fishes, as equalizers of flood flows, and as areas of mineral retention and release.

The accumulation of toxic, long-lived pollutants from our wastes—heavy metals, complex organic compounds, and other materials like asbestos—have less obvious but perhaps more serious consequences. Though present only in minute amounts, if at all, in natural environments, they can gradually accumulate to pro-

duce irreversible changes in the productivity or the adaptability of plants and animals in both rural and urban communities.

Movement of People and Materials

Urban design is closely linked with urban transportation and interventions in the transportation system; the urban planner, therefore, must begin structuring the city by consciously designing transportation and other service systems, rather than allowing unregulated growth to which he reacts. Mass transportation between and within cities, for instance, may not be self-supporting initially, but inexpensive, high speed travel between large cities and their satellites will result in significant changes in urban growth patterns. The urban planner can, with foresight, anticipate and influence these changes.

ECOSYSTEMS AND URBAN SYSTEMS

Through most of his history, man has functioned quite successfully with only a fragmented view of the world. In a less complex society with its limited population and primitive technology, he could perceive problems and conceive solutions without regard for other factors. Man's impact on the environment was limited.

Clearly, times have changed. Increasingly, the technologies we are developing to support a massive and growing human population are capable of greatly influencing the direction of geological and ecological processes. Such unplanned and unanticipated effects have had, not surprisingly, a largely negative impact on environmental stability. Obviously, if we hope to reverse these dangerous trends, or more optimistically, to manipulate environmental processes to our advantage in a sophisticated way, we must adopt a new perspective on the environment; we must recognize the "interconnectedness" of things.

The Systems Approach

The concept of "systems" arose of necessity in recent years as a new way of perceiving the complexity of our natural and cultural environments. Developed in a variety of disciplines ranging from physiology (von Bertalanffy, 1952, 1968) to operations research (Forrester, 1961), the systems concept is useful as a unifying theory for interdisciplinary communication and research. Its major strength, in fact, is the common perspective it provides in the approach to complex problems.

A system can be defined simply as elements which share one or more relationships to form a set of interacting components. Unfortunately, the simplicity ends there, for the behavior of the system as a whole is characteristically determined by the interactions among its component parts, and cannot be predicted from the behavior of any of the components viewed in isolation. The main emphasis in the analysis of any system, therefore, is on identifying the nature and relative strengths of the interactions among components, rather than on the component parts themselves.

For practical purposes, it is necessary to specify the "boundaries" of the system of interest, to identify important components and to describe the relationships among them. Usually, the system will be the smallest set of components whose interaction appears to explain the processes of the problem at hand; so defined, it will continue to interact with its "environment" (i.e., the "rest of the world"), though the particular "behaviors" an analyst wishes to study will still be governed largely by the internal characteristics of the specific system he has described. Thus, defining "the

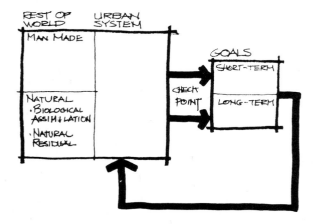

system" from a problem-solving point of view is not entirely an arbitrary process, but is determined pragmatically by the nature of the specific problem.

Our knowledge of the isolated problem, of course, will always be incomplete, especially at the beginning of an analysis. We may find it necessary to change the boundaries of the system as our understanding of it increases. Our view of the "relevant system", thus, should respond as quickly to changing *perceptions* of "reality" as it does to confirmed changes in the state of that reality. Systems science, admittedly only one way of looking at the world, seems a valuable tool for understanding the increasingly complex "metaproblems" (Chevalier, 1969) now confronting society.

The Ecosystem

"Ecosystem" is the term used to describe a particular level of organization in the organic/natural world. *Complete* ecosystems contain a sufficiently diverse set of living and non-living components to be entirely self-sustaining. A basic unit for study in ecology (Evans, 1956), the ecosystem represents a model of organization for a systems analysis of man/environment interaction.

Plants and animals constitute the living components of any ecosystem (see Odum, 1971, Chapters 2, 3 and 4) and are intimately bound together through "food chains", complex "food webs" and related processes. Green plants alone can capture the sun's energy and synthesize complex organic compounds from simple inorganic substances taken from the air and soil; accordingly, all other organisms, including man, are ultimately dependent upon green plants for their own carbohydrates, fats and proteins. The relationship is not one-sided, however, since the metabolic processes of the "consumer" organisms in turn break down the complex substances, returning simple organic and inorganic compounds to the environment for reuse by plant "producers". Plants, for example, absorb carbon dioxide and water and release oxygen in the photosynthesis of sugars; animals utilize the oxygen and metabolize the sugar, releasing carbon dioxide and water back into the environment.

Ecosystems are characterized by flows of energy and movements of matter on cyclic pathways. The well-studied pathways of nitrogen and phosphorus are known as "nutrient cycles". With such ecosystem cycles, there is no real "waste" production; very little matter is lost or gained in a balanced system. The ecosystem, in this sense, is a model for resource conservation in our socio-economic system.

Cycling, whether man-made or natural, takes place on a global scale, since all ecosystems are united by air, water and soil; global distribution of many toxic synthetic compounds is partially explained by materials recycling in natural systems.

An "external" source of energy—the sun—is required to drive the material flows in the ecosystem. However, unlike material resources, energy follows simple "linear" pathways through the system and is gradually dissipated to the environment as metabolic heat, obeying the second law of thermodynamics. The parallel to our industrial society driven by fossil fuels is only too obvious. The sun represents, for all practical purposes, a constant and infinite source of ecosystem energy. The reliability and longevity of existing fossil fuel supplies, on the other hand, is increasingly uncertain; likewise atomic power appears unsuitable as the sole answer to the rising energy demands of modern society. These twin problems of material and energy resources are a recurring theme of this report.

If we are to understand fully the natural environment, we must understand the concept of "control" as it applies to the "regulation" of ecosystem processes. Individual organisms regulate their own vital functions. Warm-blooded animals, for example, maintain a nearly constant body temperature. Such control, or resistance to change, is generally referred to as "homeostasis" and is the result of a relatively simple biological "negative feedback" system similar to the thermostatic control of room temperature.

We must understand, also, that all species have an inherent capability for continuous geometric (exponential) growth in an unlimited environment. Species' populations are potential "positive feedback" systems in which an increase in the population will result in even greater future increase. Populations of most organisms in undisturbed natural environments fluctuate, however, sometimes with great regularity between measurable limits. Clearly the very growth of these populations triggers forces tending to counteract such increases. Such "negative feedback" forces are expressed through increased death rates for a variety of extrinsic and intrinsic factors, ranging from increased predation and parasitism, through resource depletion and reduced

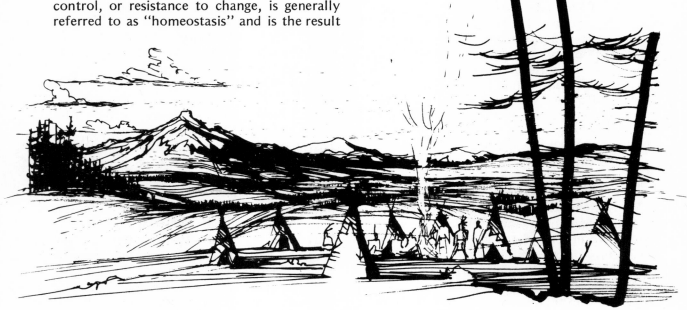

growth rates, to physiological and genetic changes.

Whole ecosystems, likewise, are characterized by self-regulating mechanisms which tend to maintain them in a state of dynamic equilibrium. Parts of such systems may oscillate regularly—for example, daily or seasonally—or irregularly, but in a "mature" ecosystem, important variables (e.g., population sizes in key species) do not change indefinitely in one direction. (There is a problem of time scale here, of course. The species composition of an ecosystem community does change gradually in sub-climax stages, as the system "evolves" toward the mature climax community. However, these are highly controlled changes that take place over a period of several years or decades. Similarly, on a geological time scale, species may come and go.)

In short, though, the important relationships between the components of ecosystems are regulated by positive and negative feedback loops; any system possessing positive feedback forces must also contain negative feedback processes, *if it is to remain stable*.

The dominance of homeostatic mechanisms or negative feedback control in natural systems is a major factor in the ability of the environment to "absorb" the impact of human activity. For example, the discharge of a limited quantity of domestic sewage into an aquatic ecosystem results in a corresponding increase in the rates of nutrient uptake by algae and decomposition of organic material by bacteria and fungi, thus maintaining environmental quality. Similarly, limited human exploitation of natural populations can result in an increase in the productive rates of remaining individuals and thus an ultimate increase in potentially exploitable adult stocks.

Ecosystems, in such ways, are capable of responding to a wide range of human interventions. The rates of processes and levels of materials or organisms may change in response to these "external" forces and internal linkages, but, in the sense that no variable continues to deviate indefinitely, the system as a whole remains in equilibrium.

There are, however, upper limits to the capacity of natural systems to compensate and maintain stability in the face of ever-increasing levels of exploitation. Phenomena such as massive algal blooms and the "sudden" decline of commercial fisheries stocks are clear indications that man has exceeded such limits, at least on a local scale. Under such conditions, ecosystem processes are temporarily dominated by excessive growth and other types of positive feedback that exceed and impair normal negative feedback mechanisms. Complete community transformation may result (Regier, 1970), "socially desirable" species being eliminated, the quality of the environment being so altered that beneficial use by man, for any purposes, is severely restricted. Ultimately, the destabilization of the global ecosystem would clearly threaten the stability and, therefore, the survival of our elaborate social systems, if not the human species itself. Human survival depends upon the emergence of strategies for human social development within the generous constraints imposed by ecosystem resilience, limits and capacities.

The Urban System

An urban system is considered by most people in terms of the "built" environment and the socio-cultural milieu man has created for himself in the city. As such, the "city" has represented the highest expression of man's technological development and cultural evolution, and by any standard is an immensely complicated entity.

Indeed, some of the obvious components of the urban system—housing developments, transportation networks, and other facilities and services, business and governmental institutions—represent extremely complex subsystems each in its own right. It even becomes debatable whether we have any truly comprehensive understanding of how cities function at higher levels of aggregation than these. Therefore, several sections of this report are devoted to more detailed analyses of the physical and institutional structures of modern urban systems, while here we look at some of the characteristics and properties of urban systems on a more general level, particularly their relationship to what we have defined as ecosystems.

Urban Systems and Ecosystems

Urban systems have many basic properties in common with ecosystems, or with almost any large and complex system. On one level, both urban systems and ecosystems are composed of many interacting components or subsystems linked together through flows of energy, matter and information of various kinds. These factors are treated in detail elsewhere; here we are interested in the similarities in the behavior of systems resulting from these linkages. Holling and Goldberg (1971) and Holling and Orians (1971) have discussed such basic "systems" properties and have related them to the historical qualities of systems, their spatial heterogeneity and the peculiar structural properties resulting from complex feedback relationships.

We can better understand urban systems by analogy with ecosystems; but it would be an

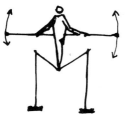
DYNAMIC BALANCE

STATIC BALANCE

CHANCE

error to superimpose all concepts derived from our understanding of natural systems onto urban systems. It is particularly instructive, for instance, to contrast the nature of the "control" mechanisms characteristic of the two systems.

Undisturbed ecosystems, we have seen, are under dominant influence of negative feedback (homeostatis) mechanisms. These dampening forces tend to move the system toward equilibrium and give it considerable resilience. Such regulation occurs between distinct boundaries defining a "domain of stability" (Holling and Goldberg, 1971) characteristic of that system. Perturbations moving the system beyond its limits result in major transformations. We must recognize, though, that the degree of internal control in ecosystems is limited largely by the intrinsic physiological capabilities, and ultimately the genetic characteristics, of the dominant species. By contrast, however, while the urban system may exhibit the same gross patterns of behavior, its basic control mechanisms are derived from the unique properties of man as a social and cultural organism.

Consequently, on the one hand, man's intellectual capacity and capability for abstract thought have enabled him to devise an array of formal and informal social and institutional controls, limited in theory only by the scope of his imagination. Many of these, such as neighborhood groups or special interest organizations, arise spontaneously from society in response to perceived threats to social stability or the status quo. Other social controls are carefully planned from the beginning. The system of "checks and balances" of the United States' Constitution is an example of the planned type of culturally derived negative feedback device.

On the other hand, our ability to manipulate our biophysical environment has led to the selective reduction or elimination of the "natural" negative feedback forces, such as starvation and disease, normally operating upon the human population. Unlike the populations of species comprising natural ecosystems, the human population on a global scale is characterized by geometric (exponential) growth resulting from the effects of dominantly positive feedback. This manipulative ability and our penchant to constantly improve the quality of human life (or at least our material standard of living) have meant that the process of material production is also characterized by positive feedback, expressed as continual economic growth. These related trends are simply manifestations of the predominantly positive feedbacks in many cultural phemonema, from the spread of panic in a mob to the advance of science and technology. Urban systems provide both the catalysts and reagents for the spontaneous chemistry of social process. The danger is that the complex process will escape human control, with disastrous consequences.

TOWARD THE URBAN ECOSYSTEM

The differences between ecosystems and urban systems will ultimately demand a synthesis of the two concepts. Historically, ecosystems have been treated merely as the "environment" of human social systems—the source of resources, the sink for wastes. Though the natural resilience of ecosystems in response to modest exploitation fostered the illusion that the environment was unlimited, we now realize that human impact on the natural environment has become a significant force in the biosphere. We are ill-prepared to confront this new reality. The positive feedback dynamics of the human economy, appropriate to an unlimited environment, are not compatible with the limited negative feedback control inherent in natural ecosystems. If we are to continue relying upon these ecosystems for our vital support functions, as we must, we require new social strategies. These can best emerge from an extended definition of the urban system.

This new conception of the urban ecosystem must go beyond our built environment and cultural milieu to incorporate the natural ecosystems upon which the human socio-economic system depends. Such a change in our perception of the "relevant" system for management and planning purposes has many important implications, especially regarding the shift from an open to a closed systems frame of reference.

By definition, the urban ecosystem will be characterized by both natural and social/institutional control mechanisms, with the latter becoming increasingly predominant. Natural ecosystem limits will largely define the scope for human manipulation of the urban ecosystem. The precise nature of social controls, however, is not predetermined. An understanding of urban ecosystem dynamics must, therefore, include an understanding of human goals and aspirations and the social and political processes reflecting and attaining them.

While we have emphasized control of the urban ecosystem through man's unique sociocultural processes, we cannot ignore the physiological and behavioral traits man shares with many other vertebrates and the basic biochemical functions and requirements he shares with most other forms of life. These biological facts of individual existence, and the constraints and opportunities they represent, can no longer be overlooked in the design and management of urban ecosystems.

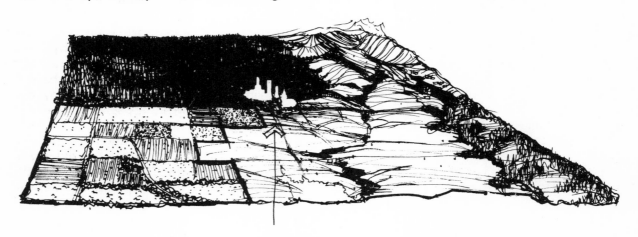

Obviously, the closed system economy of the future is a radical departure from historical trends. The transition is bound to be difficult. This reorientation has the potential to shift social concern from the quantity of production to the quality of human experience. Recognition of environmental constraints is a major step towards the liberation of humanity.

REFERENCES

Ayres, Robert U., and Allen V. Kneese. 1971. Economic and ecological effects of a stationary economy. Annual Review of Ecology and Systematics 2:1022.

Banfield, E. C. 1970. The unheavenly city. Little, Brown & Co., Boston.

Beckerman, W. 1971. Environmental policy issues: real and fictitious. Pages 19–40 in Problems of environmental economics. OECD.

Bertalanffy, Ludwig von. 1952. Problems of life: an evaluation of modern biological thought. Watts, London.

Bertalanffy, Ludwig von. 1968. General systems theory. George Braziller, New York.

Chevalier, Michael. 1969. Social science and water management: a planning strategy. Policy and Planning Branch, Canada Department of Energy, Mines and Resources, Ottawa.

Evans, Francis C. 1956. Ecosystem as the basic unit in ecology. Science 123:1127–1128.

Forrester, Jay W. 1961. Industrial dynamics. MIT Press, Cambridge, Mass.

Galbraith, J. K. 1967. The new industrial state. Houghton-Mifflin Co., Boston.

Holling, C. S., and M. A. Goldberg. 1971. Ecology and planning. Journal American Institute Planners 37:221–230.

Holling, C. S., and Gordon Orians. 1971. Toward an urban ecology. Ecological Society of America. Bulletin 52(2):2–6.

Istock, C. A. 1971. Modern environmental deterioration as a natural process. International Journal of Environmental Studies 1:151–155.

Man in the living environment. 1972. Report of the Workshop on Global Ecological Problems, The Institute of Ecology, Madison, Wisc.

Milgram, Stanley. 1970. The experience of living in cities. Science 167:1461–1468.

Odum, E. P. 1971. Fundamentals of ecology. Third edition. W. B. Saunders Co., Philadelphia, Pa.

Regier, H. A. 1970. Community transformation: some lessons from large lakes. 80th Anniversary Celebrations, University of Washington, College of Fisheries, Seattle, Washington.

Part Two

SECTION ONE:
Goals: Values and Jurisdictions

"Unless human needs and human interactions and human responses are the first consideration, the city, in any valid human sense, cannot be said to exist, for as Sophocles long ago said, "The city is people."

Lewis Mumford
Statement, Senate Subcommittee
on Executive Reorganization
April 27, 1967

The condition of the urban environment has been discussed at length in the past and is receiving considerable attention presently. Key facets of the urban environment, however, are frequently examined in isolation. One discussion, for instance, may center on air quality, another on water quality, still a third on resource flows through the city. Few examinations have attempted to view the urban environment and the urban system as a functional whole, with interdependencies and interrelations which must be understood if we are to comprehend the functioning of the entire city.

For example, the quality of the air we breathe is an important aspect of the urban system; but in considering the factors which influence air quality, we need to view not only such things as stack emissions and automobile fumes, but also the nature of the processes involved in production, the nature of urban transportation, natural air movement processes, etc. We should examine urban housing patterns, population densities and zoning codes in relation to air quality. We should look at available resources and at the technologies for converting these resources into usable energy; if coal, for instance, were to again become a primary source of residential heating because of petroleum shortages, and if the technology for converting coal into heat remains at today's crude level, our air will be polluted by sulphur dioxide and fly ash. If urban systems continue to foster private automobile transportation instead of promoting efficient, convenient and economic mass transit, air quality may decline further. If production processes use resources inefficiently or use highly polluting resources to manufacture the goods urban residents require, stack emissions may continue.

Further, the values which urban residents hold become important considerations. Action will not be taken to improve air quality unless urban residents demand it. Likewise, if urban dwellers continue to demand products that, in manufacture, are detrimental to the environment, air quality will be affected. If urban residents value green space, which can improve air quality and moderate air movement and temperature, but are unable to take appropriate action for providing suitable green space in urban areas, air quality may suffer.

The decision-making and policy-setting institutions of urban areas greatly affect not only air quality but the whole urban environment. If these institutions are not responsive to the needs and desires of urban residents, ecologically sound environmental policies will be difficult to implement. Clearly, analysis of air quality in the urban environment must examine the political processes by which decisions are made and programs implemented.

Socio-economic factors such as tax base and the amount of money urban dwellers are willing to spend to improve their air are also important factors influencing what can and will be done. Zoning standards, which determine the distribution and density of manufacturing sites, should be considered.

These and other interrelated and interdependent factors influence urban air quality. However, they cannot be evaluated without simultaneous examination of their effects on other aspects of the urban environment. If air quality is improved at the expense of water quality or landscape beauty, tradeoffs are involved and must be deliberated in the decision-making process. If, as an alternative to the incineration of wastes, land fill waste disposal is selected as a method to reduce air pollution and improve air quality, the disposal site will degrade the appearance of the land, natural drainage systems may be affected adversely, and toxic elements may percolate from the disposal site into soil or streams and rivers, and possibly into the city's water supply.

Whether waste materials are incinerated or dumped in land fill sites, reclaimable materials are lost. Many resources are becoming scarce or expensive. Burned or buried paper, for instance, cannot be recycled into usable products to reduce the demands on our forests. The reuse of glass bottles and the recycling of metal cans might reduce the among of energy required to extract and manufacture these containers. Reduction in energy use could alleviate air, water, and heat pollution in urban areas. Further, continued extraction of resources degrades the environment in regions often far removed from the city, perhaps an area used for recreation by urban residents. Energy requirements, if they are met by the construction of new hydroelectric plants, could force flooding of land which could be used for agricultural purposes; less agricultural land may reduce the food supply available to urban residents. Clearly the urban support system is far-flung, and decisions affecting urban air and water quality and waste disposal, for example, have wide-ranging ramifications.

Our discussion throughout this report is an attempt to identify, clarify, and interrelate the various aspects of the urban environment and urban system: to indicate how individual and institutional values underlie the city's functioning; to examine the nature and interrelationships of the urban system's components; to suggest interventions aimed at improving the urban environment; to detail indicators of urban system health; and to outline possible models useful in analyzing the urban system.

GOALS

We know many complex interrelations exist between and within human and natural systems; the goals and objectives of these systems are linked together, sometimes very tightly. Understanding the nature of these interrelationships, we hope to have the knowledge necessary to begin adopting an ecologically sound way of life. Formulating goals within an ecological framework is a rather novel experience; the goal-setting process is important. The choice of goals for ecological analysis in the urban system clarifies the goal-setting process and makes ecological values more explicit. We realize that ecological and social value systems can overlap and that repeating historically desired goals can reinforce their importance and relevance.

The interactions among our economic, social, political and ecological systems, ultimately, provide us with goals. These systems operate on many levels of society, the national, regional, local, or individual. Despite our desire to separate and emphasize one system or level we recognize that the holistic view and emphasis on linkage are vital. The local, regional, national and global spheres are all relevant to the full view of the urban ecosystem (Sohn, 1972).

In generating ecological goals, we must use different sets of values than have ordinarily been used in the analysis of urban areas. Ecological analysis encourages us to look at urban systems in terms of energy and materials flow, population dynamics, community analysis, and symbiotic relationships, and to employ the concepts of stability, diversity, dynamic equilibrium and trajectory. Important data will continue to come from human social and environmental relations, aspirations, and value systems. The roles and consequences of human activity are significant in the general ecosphere and more specifically in urban ecosystems. Analyzing human data in ecological terms permits us to restate human goals in more generally applicable terms.

Basic Goals for the Urban Ecosystem

Survival is the one common goal of life in general and of man in particular. This is as true for humans in urban ecosystems as it is for animals in the African savanna. Humans originally banded together and lived in groups for strength and safety. Since then, cities have been built to enhance human life beyond some minimal level of existence.

The goal of survival for the urban ecosystem, however, is disarmingly underspecified. Its identification reflects the technique of beginning a hierarchy of goals with those on which wide consensus can be reached (Berle, 1968). However, such treatment of goals does not, except in a diffuse way, meet more operational criteria—suggesting and guiding certain types of action, suggesting tools to measure and control effectiveness or recognizing internal and external constraints on achievement (Granger, 1964). Nor does it expedite the quantification of progress toward goal attainment (Terlecky, 1970).

Few, if any, proposals for improving urban areas would be entirely beneficial—if implemented without other simultaneous actions. "The morality of an act is a function of the state of the system at the time it is performed" (Hardin, 1972). The recycling of waste to conserve materials requires increased energy use. A reduction in the rate of suburban sprawl lessens the availability of inexpensive housing for the poor in the central city (Downs, 1972). Slowing the rate of economic growth increases unemployment. General goals, whether ecological or social, simply enable people to bury conflicts under deeper layers of complacency. Nevertheless, in determining goals, we have begun with the broadest. The general goal of human survival has two aspects: sustaining human life and enhancing it.

Sustaining Human Life

To state the goal of sustaining human life may seem unnecessary, but it does encourage consideration of long-range needs in an era that seems focused on immediate problems; it also provides a convenient context for relating the broad goal to the concerns of this report.

Ecologically, survival involves the successful utilization of the environment to maintain a population with all its varied activities. The city is physically structured to process materials using various forms of energy to support the human population. Social institutions are necessary to facilitate the process, to set suitable goals for its realization, and to measure the state of the system using various indicators. These components and their interrelations constitute the urban ecosystem and its functions.

The ideal for urban systems, taken from analysis of natural systems, is dynamic equilibrium in which a given supply of capital, powered by a very slowly depleting source of energy, sustains the community indefinitely. In economic terms, this is called a steady state economy. The concept, however, depends on stocks of material and energy; it indicates nothing about the social or moral state of the system. Although presently the ideal of a steady state economy is not attainable, it does provide a basis for evaluating alternative courses of action. Any policy which reduces the gap between present activities and a steady state may also help us to sustain continued human activity.

Enhancing Human Life

Merely sustaining human life is not enough; life must be worthwhile. If life remains at the level of survival only there seems but scant reason to sustain it. On the other hand, enhancing human life may indeed help to sustain it.

The ease of designating enhancement as a goal belies the enormous controversy inherent in various proposals for its achievement. What constitutes enhancement? The question has a cocophony of answers. Numerous tradeoffs must be considered—between freedom and stability, between different levels of government, between needs of the individual and needs of society, between one economic, ethnic, age or sex group and another.

In population centers as culturally diverse as

ours, there is a need to study the different individual and collective orientations with the physical environment as mediary (Klausner, 1970). By examining socio-cultural, political and economic influences on the development and functioning of human settlements, we may discover satisfying and rewarding human activities that require limited amounts of material and energy and operate on ecologically sound principles. The study of life style and energy use is particularly important in this regard (Anon., 1973, Metropolitan Bull.).

The Trends

The goals of sustaining and enhancing human life can be portrayed graphically. Figures 1 and 2 are drawn to suggest the trends that have resulted from the historical evolution of our present urban industrial civilization.

Figure 1 pictures the trend in resource utilization—our use of materials and energy. Technological advance has rapidly increased our capacity to exploit the environment. Even without increased use, energy and material resource supplies would slowly diminish. However, increased technology and increasing population have resulted in the use of resources at exponentially increasing rates. This trend can continue only for a limited period before we reach the point of no return. The curve of past experience is diverging ever more rapidly from the desired steady state

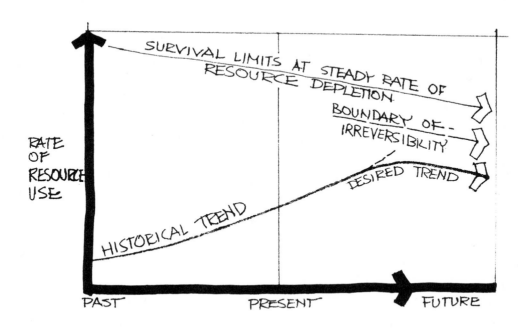

Figure 1

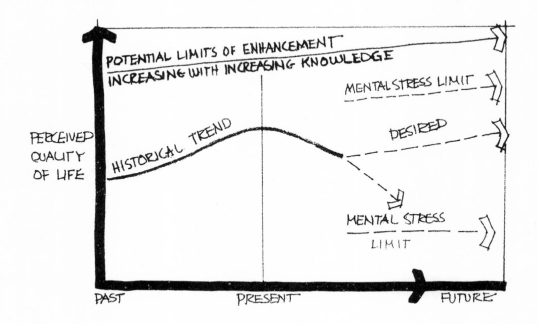

Figure 2

condition. The goal of supporting and sustaining human life can be achieved only if we move toward the steady state condition.

Figure 2 is our attempt to graphically represent the trend which relates to the enhancement of the condition of psychological man— the quality of life. This line in psychological space is controversial. It is difficult to gather consensus on whether life is getting better or worse, or whether our ability to enhance life is increasing or decreasing. We are, perhaps, more aware today of what the quality of life could be while in many respects actual living conditions are worsening. Evolution in the broadest sense may constantly be extending the upper limit for enhancement of life, increasing our future potential.

The upper and lower mental stress limits are intended to suggest that in psychological space humans operate homeostatically. The present downward trend of the curve reflects changes associated with resource depletion and increased pollution. Another trend probably could be shown moving toward the upper boundary, indicating an overload condition caused excessive attractive stimuli (Lipowski, 1970).

The two figures may suggest the outermost limits within which human life can be supported and enhanced; the perturbations caused by dynamic interactions as human activities increase are incorporated into the trend lines partially, but not completely. Presently, true limits are matters of probability and are not absolute measurements. The vagueness of the boundaries, however, does not alter the fact that at some point, they are absolute. Clearly we do not want to reach that point.

CHANGING HUMAN VALUES AND BEHAVIOR

"Far before the physical apocalypse arrives, the battle of human values will be irrevocably lost." So warn Barkley and Seckler (1972), chastising those who have overstated our ecological problems by speaking of physical catastrophe. Underlying any discussion of goals for the urban ecosystem—whether our concern is the city's institutions, its physical structure and function, its resource flows, or its populations—is the question of values. Reducing energy consumption and resource flows, for instance, may require a significant change in our values. The study and implementation of different value systems thus become key considerations in sustaining and enhancing human life.

Our apparent addiction to unbounded choice is one value especially in need of change. Lipowski (1970) has pointed out the detrimental effects of our affluence: an excess of attractive stimuli. Erik Erikson[1] says America's concern with abundance, newness and change might now come to an identity crisis. Change and newness "are in some circles on their way to becoming a matter of obsession, as if more defined identity were only too much conformity, and the only absolute choice were freedom." The situation is producing increasing numbers of aberrations as individuals do not *choose* but merely seek to experience anything. The need for structure is clear and has become a central issue in recent discussions of values and ethics for survival (Hardin, 1972).

[1]*Quoted in Milwaukee Journal, June 1, 1973.*

Our valuation of open space also requires change. Americans do value open space, although our positive attitude is largely undifferentiated. The economic values attached to land lead us to regard it as either productive or nonproductive. We may profitably redefine land by recognizing those properties of open space which make it a highly differentiated and complex resource. Categorization by productive and nonproductive uses could be replaced by two more meaningful categories, generative and regenerative. Such a view will help to insure balance between equally important uses of land and will encourage more quantification of the two.

A systematic approach is needed to determine the regenerative functions of open space and to funnel this information into environmental education programs where values can be influenced. Regenerative land functions are not solely ecological, but lie in three value categories: the ecological, the economic, and the socio-psychological.

(1) **Ecological:** Including the land's effects on micro-climate; dilution of gaseous discharge; carbon dioxide utilization; wind de-

flection; noise reduction; and water conservation and control.

(2) **Economic:** Including income generated by reduced service costs and by the value land

adds to adjacent properties; the evaluation by replacement cost at inner city locations or the irreplaceability of natural areas; and use taxes and user fees.

(3) **Socio-psychological:** Including the psychological and social values of open space: to be used as a part of everyday life (Jacobs, 1961; Newman, 1973); to enhance knowledge

of inhabitants and natural events (cf. Fried and Gleicher, 1962); to extend the range of active recreation spaces (J. Smith, 1972); to reduce the threat of overcrowding (Carson, 1972); to increase social interactions in natural areas (Cheek, 1972); and to substitute psychologically desirable wildlife for less desirable species.

Integrative studies of these three categories and of the tradeoffs involved in considering one over the others must be accompanied by policy and implementation studies. When new policy is followed by action affecting the environment, that action should be analyzed and evaluated.

Actions, when taken on the basis of a research and evaluation program allowing appropriate assessment of both the social and the environmental effects, can help insure lasting benefits. Such actions may also be useful for restructuring of our values regarding other sectors of the urban ecosystem.

Values, as we discuss them, refer to appropriate modes of conduct that will lead a person, ultimately, to some desired end state. These values are reflected in, and can be measured by, consistent behavior over time. Rokeach (1968, 1970), who has studied people's expressed values, labels the modes of conduct as "instrumental" values, and the desired end states as "terminal" values. Responsibility and ambition are examples of instrumental values; family security and freedom, of terminal values.

A value implies that a particular mode of conduct or end state is preferred over others. In actuality, the behavior of individuals and groups in complex situations reflects the operation of many value orientations simultaneously; further, many different types of behavior may be consistent with only a single value orientation. As a result, approaching consistent behavior over time as a measure of personal or organizational values is subject to interpretation, and can become very complex. Despite the complexity and difficulty, however, it is urgent that we study value systems and value change and seek to implement those values which will be expressed in behavior leading to ecological stability.

Knowing how individual, group, and institutional values are formulated and evolved and how they interact is essential to urban ecosystem studies. Social institutions—familial, cultural, political, legal and economic—operate partially independent of individual values. Personal values direct the behavior of individuals within an institution but only indirectly do they affect the operation of the institution itself. Conflicts may arise. In an organization (family, firm or agency), there is potential for three-way conflict in values between individual, organization, and society. With each layer of organizational structure, the goals and behavior of an institution become increasingly isolated from the total pattern of values within the society, as well as from individual patterns. Studying such conflicts helps determine whether or not they are resolved in favor of environmentally sound behavior by both individuals and institutions.

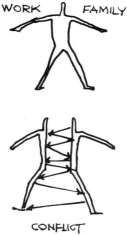

mental education programs at all age levels is essential. Many such programs are now in operation (Carson, 1974), but increased effort and support on a broader base should be encouraged.

An attractive aspect of community environmental education is its less abstract and more practical orientation, in contrast to more formal programs. The practical aspects of an involved community educating itself and implementing ecologically sound decisions are encouraging; it remains clear, however, that before this kind of program will be fully effective, both community and institutional values must change.

It is commonly assumed that once changes in zoning patterns or public health practices are made, usually above the community level, a change in attitudes or a process of "social adaptation" will follow to produce the neces-

Collectives generally have much greater power than individuals (Klausner, 1970). Resolutions of conflict between individuals and collectives does not generally lead to the best environmental results. Study of this asymmetry in specific environmental situations should receive high priority.

Values are formed very early in life through informal education. Formal education, beginning later in life, generally has less effect on values; nevertheless, environmental values can be passed on through formal education and with concerted effort the effect can be considerable. Universal environmental education seems necessary if we are to establish values leading to the acceptance and practice of ecologically sound behavior. Sound modification and reconstruction of our institutions depends on the adoption of such values. Support for both formal and informal environ-

sary support for these actions. Although many admit the fallacy of this assumption, they persist in using traditional planning strategies and policy formulations. Community or university-based leadership training and community education programs should be encouraged, therefore, to involve neighborhood and civic representatives and specialists in regional environmental planning and management (McAllister, 1973). To be effective, the coordinated and comprehensive use of physical, biological and human resources to improve the quality of life in a community depends upon communication, education, maximum participation and support. Those participating must assume a realistic view of the future. With full environmental education programs, political decisions on environmental issues are more likely to be made by groups with a keen sense of environmental limitations and with values based on sound ecological principles.

INSTITUTIONS

The institutions of our society, and in the case of our cities particularly our urban institutions, have a profound effect on the process of defining and shaping our attitudes, values and goals, and on the implementation of economically, socially and ecologically desirable programs. Therefore, before we consider the city's physical structure and function, its resource flows, and its population processes, it seems necessary to examine the urban institutions which have such wide-ranging influence. An understanding of the nature and functioning of these institutions will give insight into the process of setting goals, formulating policies and implementing appropriate programs. Further, since our institutions are such an important part of the urban ecosystem, understanding them should also reveal problems facing the city and point to possible solutions.

Institutional Progress

Institutions incorporate into their structure the results of certain decisions made at higher levels and designed to guide their day-to-day operations as they respond to changes in their environment. However, the formal institution considers only those dimensions of its environment relevant to its particular form and purpose.

In an institution, members receive incentives to follow rules serving the institution's central purpose. Again, the purpose is designed by some other segment of society with one or a few limited goals in mind.

Formal institutions are rarely responsive to the multi-dimensional changes common to complex urban ecosystems because their service-delivery or production-oriented decisions purposely exclude many environmental variables. With access to information resources such as those presumably available to an institutional policy maker, any individual observing his environment might conceivably have a clearer perspective than the institution.

To use the capabilities of many individuals to find, communicate, store and process needed information, human hierarchies have been built into formal institutions. Persons in lower eschelons are usually required to focus on specialized areas, while creative decision-making based on broader insights, intuition and judgment is reserved for a few top policy makers. Such an institutional structure can have dehumanizing effects on persons who remain at fixed levels in relatively permanent organizations. As a result of the communications revolution, however, less formal institutions are arising, often more humanly structured and more aware of ecological principles.

Informal institutions may be characterized as the more disordered, formal institutions as the more ordered. Members of formal institutions are usually employed full time on a relatively permanent basis in organizations which have, above all else, self-preservation instincts. Looser, more informal organizations—action coalitions or learning alliances, for example—tend to be more temporary and do not necessarily involve their members full-time.

"Institutional progress", a shift from the more manipulative formal institutions to more informal ones, can be considered socially valuable. Our present society may well be too institution-centered, too dependent upon formal institutions having adverse effects on nature and human nature. Accordingly, we may need to shift our emphasis toward building informal institutions.

A shift toward informal institutions represents institutional progress because: 1) it may be possible to achieve a higher degree of order or coordination from a lesser number of formal institutions; and 2) coordination through formal institutions (though they represent a higher degree of order) may be less valuable to society because of their counterproductive influences upon nature and human nature. These assertions are reflected in the following observations about participatory planning.

1. Institutions. Present human behavior is largely conditioned by artificial incentives and by acquired needs imposed through institutions having special purposes and lacking perspective. Although interdependence has increased, competitive attitudes and an emphasis on the vertical flow of products and information through institutional hierarchies have reduced the horizontal interaction among individuals and between them and their natural environment.

2. Alliances. New interactive techniques—through media, data, models, etc.—would make it possible to achieve some degree of societal coordination through temporary alliances as well as through permanent institutions. These channels should restore person-to-person and person-to-nature interaction and a sense of community and personal responsibility.

3. Planning. Planning is currently overly institution-centered. Today personal decisions about life styles could have greater effect on the urban ecosystem than institutional decisions. The chief constraints on official policies at present are the societal attitudes established during the industrial era.

4. Learning. Communications developments and, particularly, new interactive technology could help change institution-centered, production-oriented and competition-minded attitudes by freeing individuals from the artificial hierarchies fostering those attitudes and enabling them to better interact with each other and with their environment. Such interactive learning should foster more ecologically sound lifestyles.

5. Transition. New emphasis must be given to planning based upon alternative futures for individuals in a more pluralistic society. Planners in society can play a constructive role in creating options by facilitating community dialogue based on unbiased expert views buffered from vested interests.

6. Techniques. Techniques for participatory planning could include the use of interactive media, periodic issue balloting, interactive data bases, information on institutions and their activities, and community cablevision. Networks for community information exchange should incorporate these techniques; impetus for their development may be stimulated by America's bicentennial celebration.

7. Priorities. In order of priority, we need

to: translate the jargon of the planner and social scientist into information the public can understand and use; start building a community network from the bottom up; do research on citizen involvement outside academia, and perhaps construct some normative and descriptive theories inside academia.

EXISTING INSTITUTIONS

The past decade has witnessed a tremendous increase in the number, type and size of public agencies and private organizations dealing with urban ecosystem problems. In most large urban areas a complex set of public institutions has developed, partly out of local and regional initiatives and incentives, and partly out of direct actions and mandates from state and federal governments. Further, the courts have been used extensively to resolve conflicts over environmental issues and, in many cases, have given the public greater control over actions which could adversely affect urban ecosystems.

The number of private organizations concerned with ecological issues has also grown. These groups have followed a variety of structural models: some concentrate on interacting directly with the federal government; others serve primarily as resource groups, providing information and support to local organizations on an ad hoc basis; a third type of national group sometimes actually intervenes on a local, state or national scale. Some private groups are organized with divisions at all three levels; others are active only in a particular neighborhood, city or urban area.

Private groups and industries, through involvement in the legislative process, have become major participants in formulating environmental policies and helping determine how such policies are administered. Many firms have acquired some expertise in order to represent their interests in discussions on environmental questions with local, state and federal agencies and the courts, and to make, as required by law, environmental impact assessments of their activities. As a result, businesses sometimes have more skill and more information available than the governmental agencies with whom they deal.

However, partly because of the dynamic state of policy formation, and because of administrative practices and organizational behavior over the last decade, there is little understanding of how such government "public interest" and private groups are organized or how they function in specific urban regions. Partly, too, emphasis has been placed on stop-gap solutions rather than on careful evaluation of the overall character of urban ecosystems and of the mix of institutional and non-institutional strategies that could best maintain and enhance them. The tendency has often been to move from one policy to another or to give responsibility to one agency, then another, without fully understanding why the previous arrangement was unsatisfactory. Research has tended to focus on single agencies or single environmental problem areas. No one, for instance, has attempted to assess the systematic effects of all federal programs and policies affecting urban development in general or a specific metropolitan area.

Regional Boundaries and Institutional Arrangements

When viewing urban problems and institutional arrangements, we tend to think automatically that all solutions require governmental action and that action taken by more encompassing government organizations will insure better ecological conditions. Available evidence, however, indicates that neither assumption is necessarily correct; or if either is, it is only under certain conditions. Indeed, one may argue that "public interest" groups and citizen activities have been more effective in protecting urban ecosystems than the responsible national, state, and local government agencies.

Two of the common difficulties besetting governmental institutions have been: 1) narrowness of concept and a lack of understanding about the interdependence of the elements of the urban ecosystem; and 2) lack of jurisdiction as large in concern or responsibility as the problem areas being addressed.

Broadening areas of jurisdiction to solve urban problems would require an understanding of the interdependence of all units of government and the matching of their responsibilities with the capacity to act. Non-governmental institutions may have a powerful role in educating public officials.

In the United States, the federal government has the basic responsibility for carrying out the broad national purpose; it financially supports subsidiary units, but can be faulted for piecemeal and often narrow approaches to regional or local problems and for little understanding of how individual program actions may affect other elements in an urban ecosystem.

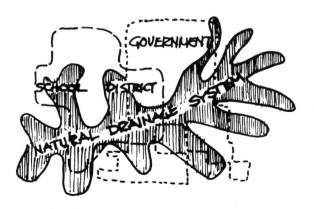

At the local level, jurisdiction is often too restricted to permit constructive action on broad aspects of urban ecosystem elements. Local governments could play a substantial role if their actions and responsibilities were part of a well-planned and integrated institutional system interrelating federal, state, regional and local governmental responsibilities and capabilities based on clearly defined principles.

These two points are important: comprehensive planning, at all levels of government, must be based on an understanding of ecosystem principles; a planning and action system interrelating government functions and corresponding responsibilities at various levels is necessary.

For some purposes, of course, a city, defined by its corporate boundaries, must be treated as a closed system. But an urban area usually encompasses several cities, special authorities and counties that are highly interdependent; its governmental units, in combination, constitute its public institutional structure. Any findings concerning the ecology of a single city, to be fully useful, must be placed in the context of the larger ecosystem and governmental system; we find, however, several obstacles to achieving this goal. At the local level, for instance, urban officials generally have no cognitive framework for considering the area ecosystem. Environmental agencies or advisory councils, if established, are frequently without adequate funds or staff and sometimes without influence as well. Public interest groups are often the only ones that view the total area in ecological terms, but they frequently have a limited set of values, making it difficult for them to deal with unavoidable tradeoffs and compromises. Further, policies and agencies are formulated or developed to address only one problem at a time, and citizen groups, because of limited resources, often have difficulty maintaining active campaigns on more than one issue at a time. What results are fragmented views of area ecosystem problems.

Within the agencies that do exist, moreover, there are numerous operational problems: there is little emphasis on cross-fertilization between units or agencies dealing with similar problems or operating within the same general region; the agencies often focus on physical and economic rather than social considerations; there is little development of a professional staff with the technical capacity to deal with environmental problems; there is little money available for staff or research, so agencies are dependent on other groups for information and explanations and face difficulties interpreting what information they do receive; the decision-makers within the agencies are often appointed and are not directly responsible to the public.

Within urban regions or large cities, there is currently a trend toward, or at least interest in, putting the decision-making powers into the hands of the affected subareas. There have been, however, no efforts to relate these moves to regionalization and decentralization of environment-related responsibilities. There has been, also, a major trend toward the development of public interest groups to represent ecologically-oriented values before legislators and other decision-makers; these groups are seeking the reorganization of formal structures and are attempting to inform the public about ecological matters. Nevertheless, there are still few mechanisms at the regional or local level for providing the public with adequate and unbiased ecological information and for detailing the implications of proposed public and private projects. There has been little effort to facilitate public participation in making decisions on ecological problems or to evaluate and make reasonable tradeoffs between conflicting values.

Attempts to define "regional" and "institutional" boundaries and to correlate the two have faced numerous difficulties. Boundaries of municipal governments have tended to grow larger as urban areas have grown in size and density. Sometimes, additional cities have even been organized within larger urban regions. As the population size and boundaries of the city enlarge, so do the boundaries of the physical subsystems needed to provide resources and/or affected by urban processes.

A number of efforts have been made to define boundaries for a single governing unit that would have jurisdiction over the physical, social and economic subsystems of a metropolitan area. The scale of these subsystems however, is generally larger than that of the municipal governing units and is organized differently.

Efforts to divide control over the urban system, putting appropriate responsibilities at local, regional, state and federal levels have also faced several difficulties. As social, economic and technological processes change, for instance, they no longer fit current problems. Again, influence and effective control do not always operate in the directions intended. State agencies may dominate the policies and operations of a federal agency; local agencies may do likewise in relation to state or regional authorities.

And there are other problems with large municipal or region-wide governments. Large-scale units, in their decision-making, tend to be insensitive to the conditions and preferences of their small subareas. Citizens often have problems both in obtaining information about public and private actions affecting them and their section of the city and in exercising influence in the larger decision-making process.

Careful consideration, therefore, ought to be given to strategies which:

1. try to concentrate as much control as

possible over ecosystem policies at the lowest level of government, or to design large-scale units so they focus attention on these sub-areas;

2. put emphasis on mechanisms among governmental units to regulate those factors affecting the urban area but lying outside its political boundaries, rather than on establishing large-scale units to provide public goods and services directly;

3. create mechanisms within large-scale units of government that will provide the public with adequate information about the current state of the ecosystem and the implications of significant public and private projects in their area, facilitate public participating in governmental decision-making, and evaluate trade-offs among conflicting values pertaining to the state of the ecosystem;

4. facilitate citizen organizations that provide for a non-institutional means of maintaining and enhancing urban ecosystems; and

5. establish strong links between interdependent levels of organization—from the neighborhood to the national level—to maximize the flow of information.

Control mechanisms and decision-making processes should be visible, identifiable, readily accessible, and politically responsible.

The Impact of Information

Because fragmented views of the urban ecosystem are detrimental, the urgent question arises: how can information about urban ecosystems be disseminated and used to influence management and performance? We assume changes in institutional responses to urban ecosystem problems will result from some directed action taken by individuals, informal groups or formal institutions. But, in order to act, they must have reliable information about current conditions and they must judge the situation as right or wrong according to

their own values and perceptions. When the group's perception of the real world is in harmony with its value system, no action will be taken. There are indications also that available information and knowledge of causal relationships are not always sufficient to bring about action. Those receiving information often avoid action by rationalizing the differences between their values and their perception of the real world.

If action is to be stimulated, strategies must be developed to determine the potential impact of information. What results may be the exercise of institutional responsibility; pressure to change institutional authority or responsiblity; the formation of informal pressure groups; or individual responses. Accordingly, several issues need to be investigated. One refers to citizen access to information possessed by public and private institutions; questions include the form of the information, its availability and timing, rules governing its disclosure, the effect its release will have on the institution's objectives, its relevancy to ecological concerns, and alternative methods of acquiring it if not readily available. Another issue concerns the use of available information. How, for instance, does one acquire the technical expertise to interpret information, and what can be considered a significant difference between the reality of a situation and the values held by a group or individual? As these issues are clarified, governmental and public groups and individuals will be able to take more rational action to change the operations of existing institutions and perhaps even change the institutions themselves.

Factors Underlying Institutional Change

The capabilities of social, political and economic institutions, in general, are constrained by their area of responsibility and authority, their scale, resources, links with other agencies, their recognition of and response to citizen concerns and their ability to accommodate to changing social values. The success of any institutional change depends upon the interplay of these factors; we cannot assume that change in a single factor will produce the desired results. As these factors are applied to the assessment and management of urban ecosystems, several research questions may be asked:

1. What is the match or mismatch of management areas and general political units?
2. How are natural systems matched with urban political bodies?
3. How is funding for urban ecosystem controls matched or mismatched to responsibility?
4. To what extent is there a capability to generate and use information among and between interested parties and regulating agencies?
5. Are there unalterable or nearly unalterable spatial and functional patterns naturally present in urban ecosystems and can institutional arrangements be devised to match these patterns?

NEW INSTITUTIONS

Alteration of institutions not responding satisfactorily to environmental problems is one obvious way to correct them. Since many institutional changes have had unintended and often negative consequences, however, there must be some indication that a new institutional structure will achieve the desired response. Institutional change, if it is desired, is often easiest to achieve in times of crisis—but one cannot always rely upon the convenient occurrence of a crisis. Change can also be induced through the shaping of political awareness by introducing new information which will change the perception of the decision-maker. One of the major weaknesses of "urban environmentalists" is their inability to substantiate the relationship between existing institutional arrangements and institutional failure to produce effective urban environmental policy. They employ a good deal of rhetoric, much conventional wisdom, some sound assertions, but little effective evidence. The state of the art is primitive, and what is known is poorly communicated to the political practitioner.

Determining how different structural arrangements will produce different policy outcomes, therefore, is a primary consideration for those proposing institutional re-engineering. The relationship between structure and policy, in fact, is one of the major research tasks facing political scientists today. A second consideration in institutional development concerns the way to effect institutional change; efforts at reform are often doomed simply because their architects misread political realities or use wrong strategies.

A third consideration is the question of institutional obsolescence—the adequacy of the present organizational theory of institutions and the effect of present divisions between governmental and non-governmental institutions. Hierarchical organizations, whether they are in government, industrial corporations or the university, are a product of the industrial age; they are highly efficient at turning out certain products, but they can also be severely limited in dealing with multi-dimensional problems and can have a dehumanizing effect on those working within them. They are product-oriented rather than process-oriented. There are, indeed, increasing signs that one of the factors hindering effective treatment of the urban ecosystem is not a particular set of institutions but the unsuitability of the hierarchical structure of those institutions.

One alternative is the "collegial" organization or horizontal decision-making system. Transportation decisions, for instance, would not be made by a department of highways or a high government official, but by mutual agreement of government officials, politicians, professionals and affected citizens. There should be experimentation with such forms of decision-making utilizing a coalition of decision-makers and benefitting from the multiple views involved. We are not suggesting that all forms of hierarchical government be dissolved, but that a mixture of different forms be developed. Similarly, the distinctions between public and private, or governmental and non-governmental, activities need to be redefined; government intervention in the private sphere, even when undertaken with the best of intentions, is often counterproductive.

Principles of New Institutions

Our existing institutions are not sufficiently responsive to urban ecological issues. New institutions, therefore, must be founded on the principle that they be responsive to pressures related to all elements of the urban ecosystem; to be responsive, they must first be able to receive information from the various subsystems, then be able to act in response to it. There are institutions which appear to be developing such capability. We will discuss some of these and will suggest research and monitoring activities to increase our understanding and to encourage employment of such responsive institutional structures.

We can identify, within any given population, groups having a wide array of environmental perceptions. Such diversity can be attributed partly to the variety of cultural groups within the population, and partly to the variety of physical and biological environments surrounding the population. To recognize the full range of perceptions within a population, an institution must recognize first this environmental and cultural diversity.

Social structures are determined largely by communications systems and by other variables related to communication. For instance, with the advent of high density living, automobiles, radio and television, the traditional neighborhood units have been replaced, in many cases, by special interest groups. Recognizing and focusing attention on these groups can facilitate the communication process and help groups identify and make use of the expertise available to them.

Additionally, the term "job" should be freed from its production focus and from its role in the distribution of income; people must be allowed more freedom to find and then implement solutions to the city's problems. Further, the institutions facilitating these efforts need to become less specialized and to redefine their jurisdictional boundaries to conform more closely with the ecosystem they are influencing. The fragmented, disciplinary approach of many current institutions, whether industrial or governmental, and the unnatural boundaries of their areas of influence, have hampered solution to, if not created, the problems we now face.

There are only a few examples of attempts to match the degree of political decentralization with social and environmental patterns. Two are in Canada. One is the city of Winnipeg, which has devised a system of local government consisting of two principle parts. One part is the system of community councils and the city council which responds to clearly defined, operating problems within the urban ecosystem and which has the decision-making function. The second part is a more informal system of resident advisory groups whose representatives are elected from public meetings to accommodate issue and interest groups within the community. Working together, these two groups represent an experiment in political decentralization that deserves attention.

The second example exists on Gabriola Island in British Columbia, where a small population residing in an idyllic recreational area is attempting to stave off the loss of their environment to suburban development.

Canada is also experimenting in breaking up the close links between income distribution, production of material goods and jobs. The most notable of these have been the "Opportunities for Youth" (OFY) and "Local Incentive Programme" (LIP) made by the federal government. These programs were conceived and implemented to combat high levels of unemployment, particularly among youger age groups. They have permitted new constituencies to emerge, many of which are self-supporting. The programs have, in fact, enabled several "environmental" groups across the country to mobilize human resources, previously untapped, to help search out and implement solutions to urban problems.

Some experiments in the regionalization of administrative units have also been undertaken. The Winnipeg example is one. Another was made possible by a section of the Municipal Act of the Province of British Columbia. This Act allows a regional administration of urban systems to evolve slowly. As municipalities conclude they need larger or smaller administrative units to cope with their problems, they can adopt a more suitable administrative structure. Similar experiments exist elsewhere in Canada and in the United States. These experimental units are evidence of evolutionary trends in the administrative structures used to manage the urban ecosystem. The creation and encouragement of other experimental institutional structures is the only way to examine the full range of possible institutional arrangements and to assess their potential for facilitating interdisciplinary approaches to ecologically sound management.

Centralization and Decentralization of Authority

Two problems plague present efforts to reform governmental institutions. The first need is to overcome the fragmented authority of agencies in the urban area and to reorganize decision-making units on a regional scale to better match powers and problems. The second need is to decentralize decision-making so individual citizens feel in touch with government. Several urban areas are testing new regional institutions—the Regional Planning Commission in San Diego and the Regional Council System in Ontario are examples. There have also been efforts to decentralize government, such as the development of neighborhood city halls, community development corporations or citizen advisory groups. It may appear, on the surface, that making government both regional and decentralized is an unworkable, if not impossible, task; but the two objectives can be brought into harmony.

The new government of the Winnipeg system, mentioned earlier, was organized with both principles in mind. It brought the central city of Winnipeg and thirteen suburban municipalities together under one administrative system with integrated services, planning processes and tax roles. As a result, there is one general-purpose regional government with one council and one administrative system.

At the same time, the Winnipeg political system was decentralized and efforts were made to increase citizen involvement. The city is now divided into 50 wards, each with 10,000 voters; these small constituencies enhance contact between citizens and their elected representatives. Three or four wards are brought together in a unit called the community committee, which has powers over land use, service delivery, parks and recreation. Three or four councilmen on a community committee act, in effect, as a subcommittee of the full city council. Attached to each community committee is a resident advisory committee composed of about 40 local residents chosen once a year at a community conference.

Initial assessments indicate that resident advisors are beginning to play an effective role in decision-making, not only advising elected representatives, but in many instances becoming the actual decision-makers on local issues. They also respond on a regional basis, since the city council refers many issues to the resident advisory groups.

The Winnipeg experiment, then, appears to be one successful way of marrying the concepts of regional government and decentralization, although it still remains to be seen whether the model can be applied to larger urban areas.

The New Jersey Development Corporation is another illustration of such an approach. Until a change in party control in 1968, the New Jersey government was planning to introduce a system of development corporations. A state urban development corporation was to set general priorities for development, new town policy, etc., on a state level. A network of local development corporations, linked to the state group and based on citizen support, could call on state agencies for technical assistance and powers to carry out their duties. Each local group was to be composed of politicians, citizens and administrators. The use of this concept to integrate local participation and broad development or management efforts should be explored further.

A third example of the concept is the Canadian Tri-level System, designed to circumvent the problem of having externally-made decisions impinge upon local urban areas. In each major city, a tri-level committee is organized, composed of local, provincial and federal officials; they discuss common problems, organize common responses to them, and try to reduce any conflicts between different governmental programs. The tri-level committees, for instance, are discussing where airports should be located and how federal urban land use can be made to match local land use plans. The system, though still not fully developed, illustrates a new institutional response to the problem of fitting the scale of action to the scale of the problem.

These three examples suggest that we should examine the linking of institutions on the government level and the linking of public officials and private citizens. Closer ties may be desirable.

Professional Roles in New Institutions

One critical question regarding new institutions in the urban environment is the role of the professional. New skills are required to link citizen organizations, government and business. Several types of urban professionals have developed—storefront lawyers, architects, public advocacy professionals, for instance—and universities and community colleges have become more involved in community-related activities, serving a wide range of community groups and lending advice and expertise to decision-makers. The educational institution, in fact, can provide a home base for the new urban professional who, through activities in the community, can demonstrate an increasingly important advisory role for the university.

A summary of new professional activities would include: public interest research; university outreach programs; information resource centers; and community professional projects such as those of the community design centers in New York State.

There are, however, two sets of problems associated with the establishment of a system of urban professionals who could enhance policy-making on urban environmental matters. The first is training; the second is tentative support. Most professional skills have not been geared towards the necessary interdisciplinary and community-related training. Once such professionals are trained, they exist precariously; there are few university urban programs or community design centers, and those in existence are not entirely stable. We must accept the need for a new kind of urban professional and provide effective training programs for him; at the same time, we must gain public support for programs or centers within which these professionals can perform their tasks.

Institutional Change

Cultural institutions have been among the key resources that have made cities centers of civilization. These institutions, often located in the urban core, now face severe stress as middle and upper income groups move to the suburbs and have less frequent contact with them. The institutions' new neighbors are often immigrants to the city who cannot penetrate the institutions' rigid cultural barriers. Yet the cultural institutions must function, as they have in the past, to enhance the lives of the new urban dwellers and to keep suburbanites mindful of what urban areas have to offer.

The physical resources of these institutions cannot easily be transferred, nor can they be duplicated in new suburban areas without extraordinary cost. Most of these cultural institutions thus find themselves in severe financial straits as operating costs rise and financial resources decline; few have the resources to alleviate the condition.

Recognizing the importance of these institutions in enhancing the quality of urban life, several cities have taken steps to enable their institutions to work together to develop programs responsive to the changing urban population. Philadelphia and Buffalo, for example, are both enhancing the educational and financial capacity of their existing cultural institutions, as well as encouraging new institutions which will better serve each city's new residents.

BANC, "Build a New City"—an association of cultural institutions in Buffalo—grew from the efforts of a citizen task force attempting to revitalize a central city park for the sesquicentennial birthday of landscape architect Frederick Law Omstead. All of the city's key cultural institutions, including a branch of the state university, were near the park, yet most of them were indifferent to the park beautification project because they were caught up in their own serious internal problems.

About the same time, a master plan report was released from the state university system announcing that its ten-year physical expansion, during which the university had grown to 70-odd campuses, had come to an end. Future expansion, the report indicated, would be directed toward existing community resources. This prompted the university branch in Buffalo to initiate a series of meetings with its institutional neighbors.

The board presidents and the chief administrative officers were invited to the meetings; only about six institutions were invited at any one time. A range of different institutions were deliberately brought together so that art institutions would, for the first time, communicate with hospitals and zoological societies to find that they had common interests. A transcript of each meeting was sent to the participants, as well as to institutions not attending. Most institutions responded favorably. After all had attended a group meeting, they were given a questionnaire which sought this information:

1. Describe the basic purpose of your institution, including the population to which you provide services and the nature of those services.

2. (a) List in order of importance the major problems your institution is facing.

(b) Which of these might be solved in whole or part through common action?

3. (a) Describe briefly short- and long-range institutional plans.

(b) Could you better carry out any of these plans in cooperation with other institutions?

Approximately 20 institutions responded to the questionnaire. Problems listed range from academic inertia to vandalism. The questionnaire and responses to it became the focus of a large, all-institution meeting where representatives were asked to join task forces designed to explore financial, community and institutional resources. Over a period of months, the task forces met and developed concrete proposals for institutional cooperation. These included a financial management plan, a vehicular purchase and maintenance system and a proposal for institutions to look carefully at their community relations.

After reporting on these proposals at a second all-institution meeting, a steering committee was formed to develop the superstructure for this new association of cultural institutions. Retaining the name Build a New City (BANC), the group decided to incorporate and to develop an administrative staff. The task forces will now become permanent committees, and steps are being taken to implement the specific proposals for institutional cooperation.

In looking for better ways to manage urban systems, there is need for the broad involvement of formal as well as informal institutions. The concept of BANC is being communicated throughout western New York State and could serve as a model for cities seeking to strengthen or revitalize their cultural institutions.

LINKS BETWEEN INSTITUTIONS

An assumption in urban studies has long been that important decisions are made in hierarchical fashion: that information flows primarily up the hierarchy to "decision-makers", with little communication across jurisdictional boundaries. Recently this approach has been called into question, for many ecological problems require solutions which span several jurisdictions and go beyond town, city, county, state and even national boundaries. We are concerned here with developing new ways to facilitate communication across these boundaries so that ecologically sound goals can be developed and urban problems can be identified and resolved before they reach a crisis state.

We have noted previously that multi-dimensional ecosystem problems cannot be resolved within narrow-purpose institutions. New broad-purpose institutions designed to implement urban ecosystem policies should recognize free-flowing community dialogue as a necessary precedent to formal decision-making by elected officials. Such dialogue could often lead to non-institutional solutions. For example, it could inspire concerned citizens to share their cars in order to combat air pollution, urban congestion and the energy crisis. Institutional arrangements might be required to follow such an agreement—to insure safety or to provide added incentives for hitchhikers and cooperating drivers, for example. Community dialogue can facilitate coordinated citizen action, with little if any institutional support.

To encourage community dialogue, a network of community information exchanges should be established. A permanent information exchange network is needed to support and encourage temporary community alliances. The communications technology necessary to create such a network is becoming increasingly available with computers, two-way cablevision and various innovations in telephone systems and printed media. But these network technologies will not be used in community information exchange unless community nodes or ties are first established.

These ties could take a variety of forms—links between community design centers, public interest groups and information services, for example. One type of node might be a "community place" with portable video equipment to record community situations, or interactive computer graphics to display environmental indicators and provide information on institutions, their programs, products and services. It would be equipped to record citizen opinions, complaints or suggestions. Citizens could receive training or instruction in the centers to familiarize themselves with institutions, issues and experts and to encourage their dialogue with other interested persons.

Research is needed now to determine the basic design criteria for these community nodes. Perhaps by 1976 a network development program should be underway. In order to avoid jurisdictional complications, node development should be encouraged initially at the neighborhood and regional levels. Network development should focus at the continental level, and by 1980 the commitment to a global network for community information exchange should exist. This commitment will probably have to be made by the independent sectors of society—citizen corporations, non-profit organizations, foundations, cooperatives and religious organizations. The information nodes should be tailored to the needs and cultures of diverse communities and organized at the grass roots level. The global network should utilize and contribute to the Environmental Information Referral Service currently being established in Geneva, Switzerland, under the sponsorship of the United Nations Environmental Program.

Links Between Developed and Underdeveloped Urban Areas

A schism between developed and underdeveloped urban areas, similar to the international schism between developed and underdeveloped countries, exists and saps the resources and strengths of urban areas, creating enormous costs. The schism can be reduced by creating new institutions in underdeveloped urban areas which, linked to institutions in developed areas, can help reduce disparities between the two. Although the United States possesses the resources to reduce or eliminate the schism, the country has been unable to do so. The social complexities involved, i.e., racial and economic issues, have yet to be resolved.

Efforts have been made to reduce the schism by formalizing institutional structure within underdeveloped urban areas through community organizations. The power-based community organizations developed by Saul D. Alinsky in Chicago, Rochester, Buffalo and other areas are examples. Parallel to such efforts was the growth in advocacy planning that attempted to bridge the schism by creating a flow of technical information across it. Paul Davidoff, who coined the term "advocacy planning", has been a strong supporter of this approach.

The formation of community design or development centers throughout the United States can provide the technical links between the developed and underdeveloped areas of our communities. Two such centers are the Architects Committee to Renew Harlem (ARCH), formed in the early sixties, and the Community Planning Assistance Center of western New York (CAPAC), operating in Buffalo for several years. The control the communities being served have exerted over the operations of both these technical centers has been significant in their development.

The development of Alinksy-type organizations is usually stimulated by the emergence of some broad community issue drawing response from many groups. The resulting organization is generally a confederation of several existing groups or informal institutions lacking the resources, structure or connections to sustain themselves or to exert sufficient influence over the decision-making process.

The Woodlawn Organization (T.W.O.) in Chicago, for example, was formed to challenge the efforts of the University of Chicago to expand into the surrounding black community. B.U.I.L.D., "Build Unity, Independence, Liberty and Dignity", was organized in Buffalo when the communications media provoked existing groups in the black community after riots in New York and Rochester by saying that such problems did not exist in Buffalo's underdeveloped inner city. F.I.G.H.T., "Freedom, Independence, God, Honesty, Togetherness", was formed when the developed area of Rochester failed to respond to the issues raised during the riots of 1964. Each of these new community organizations has now been in existence about a decade. They have successfully evolved from politically-based power groups to ones which have a substantial economic power base as well.

The patterns for these community organizations are similar. The need was recognized for some formal structure to develop links between existing institutions concerned about an issue. Accordingly, an umbrella organization was originally formed in each area at a convention bringing the concerned groups together. Now, a convention meets each year to elect the organization's officers and a steering committee; the convention can also be called together at other times to provide input on major issues. Responsiveness on the part of the umbrella organization is insured by the yearly elections and by the accountability of the officers to the steering committee. Any organization, formal or informal, can join the umbrella group—all it need to do is pay the usually modest convention fee.

Initially, the resources for these organizations came from both internal and external sources. During the sixties religious groups and foundations were the chief contributors, while internal sources contributed only a modest amount. After consolidating their political power, the organizations were able to become self-sustaining through their economic control of new developments and programs—the ownership of nonprofit housing, for instance, or the management of a community based school or the development of new private companies.

These community organizations do not necessarily insure political responsiveness to issues affecting the underdeveloped urban areas, however; the umbrella organizations have started to initiate alliances with other groups in the urban area who share similar problems, though not necessarily the same overall philosophy.

Education, for example, is of major concern to groups in underdeveloped areas of the city who are affected by busing programs and who may feel that success lies rather in improving the neighborhood school, whatever its racial complexion. This concern has led to alliances with more organized blue collar worker groups who are also concerned about the quality of schools, but who reject integration, and with upper income families who are facing the increasing costs of private education. In the future, these improbable allies can also be expected to combine forces on other issues.

Community development projects in the past have lacked adequate internal organization and the involvement of needed professionals or technical experts. Now, however, the development of community organization and the rise of advocacy programs within our educational institutions have facilitated the emergence of new institutions which can assist underdeveloped municipal areas in acquiring needed skills and in applying them to high priority community projects. These design/development centers have loose regional structure, patterned after the ARCH program begun in the early sixties in Harlem by a group of young professionals who rejected the traditional practice of architecture and planning and went instead into the Harlem area to assist the community in developing itself.

Meanwhile similar spontaneous ventures began throughout the United States; some were the work of collections of individuals; others involved local architectural, legal or other professional firms. The Community Law Office in Harlem is one example of an advocacy project in the legal profession. The media have helped make these activities generally known.

Today over 70 community design/development centers (CDC's) exist in the U.S. The Architect's Workshop in Philadelphia and the Community Planning Assistance Center in Buffalo are two examples. In these, core administrative structures place volunteer professionals with community groups seeking their skills. Community control of the administrative organization and the selection process has been essential to the development of responsive CDC's. The Community Planning Assistance Center, in fact, requires that community representation be in the majority on their board and executive committee to prevent control by an elite group.

Most CDC's are not self-sustaining and must rely on outside resources for their support. The support for the center in Buffalo, which has no fees, comes from agencies like the New York State Council on the Arts and the local Model City Agency. Other groups have received funds from public and private agencies such as the U.S. Department of Housing and Urban Development and the Office of Economic Opportunity. The goal of all community design/development centers, of course, is to be self-sustaining. In New York state, that goal is being fostered by the development of a statewide association of CDC's; funds for the umbrella organization will come from the New York State Council on the Arts or from the New York State Urban Development Corporation.

Community design/development centers are

useful institutional structures for linking formal and informal institutions already in existence and for mobilizing their resources and energies. They can and will play significant roles in the formation of socially responsive and ecologically sound urban ecosystem policies.

ECOLOGICAL ACCOUNTING SYSTEMS

Urban institutions have significant impact on the formulation of goals and policies for the management of the urban ecosystem. We have found, generally, that existing institutions are not sufficiently responsive to ecological issues affecting the city and we have, accordingly, called for changes that will enable them to react appropriately to the city's multi-dimensional problems. We have also called for the creation of new institutions which seek to involve all the various sectors of the urban community and to respond to the issues those citizens raise. Finally, because of the informational and jurisdictional problems involved in adequately assessing and managing the urban ecosystem, we have outlined the need to forge new links between urban institutions so that policies can be formulated with all aspects of the city in mind.

Individual citizens, informal groups, and formal institutions, however, if they are to respond effectively to ecological issues in urban areas, must begin to recognize the need for a new urban accounting system. Presently society's values and goals are expressed primarily in terms of economic growth and economic accounting systems. We urgently need a more ecologically based accounting system, for example, energy and resource flow to replace dollar flow, or a phosphate standard to replace the gold standard.

Our present economic values of growth and development can produce only short-term benefits and are oriented toward individual rather than societal gain. Under this system, the individual can achieve a very high standard of living now, without altering his values. But this behavior, were it to go unchecked until resources become depleted, could result in the destruction of man and other species.

In modern man-dominated ecosystems the economic and political systems have become the primary institutions checking or constraining human behavior. Humanistic institutions have assumed a secondary role. As a result, in the United States, living conditions for inner city residents are often dismal. Our limited resources are being rapidly into goods for consumption; inefficient production of those goods contaminates the ecosystem and impedes the return of basic materials to the biosphere.

The market mechanism, i.e., pricing, fails to reflect the benefits and costs, to society, of resource transformation, and reflects instead the benefits and costs to the private parties involved. Prices are largely determined by the distribution of income and do not contain feedback mechanisms to prevent resource depletion and environmental degradation; both are very costly to society. If man is to make wise and efficient use of his resources, the market mechanism must begin to reflect societal costs and the tradeoffs between short- and long-term benefits.

Of course, changing the market mechanism, alone, is not sufficient to accomplish this task. Economic growth must also be reduced. "There is no conceivable way to produce, use and eventually dispose of economic commodities without creating some degree of environmental decay. The longer economic growth proceeds, the more severe will be the environmental decay" (Barkley and Seckler, 1972).

Changing the market mechanism and reducing economic growth both have disadvantages. Internalizing costs not currently factored into the price of a product, for instance, will hurt marginal producers and consumers. Reducing economic growth will also create inequities, probably by increasing unemployment that will have to be compensated for by increased subsidy programs and perhaps a guaranteed income plan.

Guaranteed annual income is a highly controversial issue. Yet Johnson (1971) observes that such an economic tool might contribute significantly to the much needed encouragement of labor-intensive activities and a reduction of resource- and energy-intensive activities. A policy of guaranteed annual income would require a significant value change for the United States and might ultimately lead to a ceiling on disposable income as well. In the long run, such a system would have beneficial environmental effects, but many view it as a spectre that could result only in severely limited freedom.

Clearly, if we are to begin living symbiotically in our environment and move towards approximating a steady state economy, we need to find new ways to measure the full costs of maintaining the urban ecosystem.

CONTINUOUS EVALUATION OF DEVELOPMENTS

We must begin continuous monitoring of changes taking place in our environment and use that information to guide and control future developments. Without such information, it is unlikely that ecologically sound decisions will be realized on a scale large enough to halt environmental degradation and to reduce the rapid consumption of energy and resources.

Our major tool for evaluating developments in the urban ecosystem, as we have indicated, is economic. All other aspects of development—materials, access, services, style, even political and social acceptability—are translated into economic terms. But economic indicators are insufficient for adequate evaluation of urban ecosystems. They fail to evaluate many environmental changes resulting from urban developments, and mislead because they always evaluate rapid growth in positive terms.

Ordinarily the indicators used to evaluate environmental changes, particularly environmental degradation, are the nonmarket or amenity type: pollution; reduced open green space; loss of wildlife, wilderness or unique environments; increased noise; loss of well-being. Amenities either have relatively little economic value or they are evaluated inconsistently because of divergent value systems. They are also generally noncoercive. Their preservation has been managed by institutions setting rather arbitrary standards, since data to support any standards are difficult to obtain. Those standards have typically been minimal because the complexity of urban ecosystems has been oversimplified, resulting in long-run deficits, and because there has been unequal distribution of amenities, related to but not entirely dependent on economic inequities and unequal distribution of information. Corrective measures are needed, not only to develop more adequate indicators, but also to develop and perfect the system of communication and to support action based on the information received from the monitoring.

Gross and Refined Indicators

We can distinguish between gross and refined indicators of environmental change. By a gross indicator we mean one that is immediately apparent to an untrained observer, although its meaning may not be clearly understood. Examples include agal growth in a stream or an irritation of the throat.

By a refined indicator we mean one that requires sophisticated equipment to measure conditions not immediately apparent even to the trained observer. An increase in phosphate concentration in a stream or sulphur oxides in the air are examples.

The relationship between gross and refined indicators emphasizes the need to start using more refined indicators rather than waiting for the potentially irreversible conditions revealed only by gross indicators. The basic issue is illustrated by Stainbrook (1968), using a medical analogy. Waiting for biological detectors to indicate a substantive change called lung cancer places us beyond the point at which preventive action could be effective.

Further, we need to begin communicating the state of the ecosystem as made evident by our indicators, and to encourage and ultimately enforce recommended action. Accordingly, the indicators and methods of presenting them to the public and institutional and governmental decision-makers must be examined within our legal and political systems. Initially, an adequate system for continuous monitoring and feedback of environmental change would require a large investment, but its operation would enable us to avoid far more costly irreversible conditions. Clearly, some of the resources now directed toward development have to be redirected to establish such a system.

It is important that the meaning of the information reaching institutions and individuals be accessible without laborious calculations and inferences; users should be freed of these tasks. It is also important, for the presentation of information, that the display be integrated: users must have predictive displays that allow them to know the consequences of their collective actions. Rather precise predictive models of indicators reliably measuring environmental change must be developed for use in such displays.

The emphasis on continuous monitoring and feedback of environmental changes underscores the need to shift away from *summative* evaluation—simple comparisons of before and after conditions—towards greater use of *formative* evaluation—assessment intended to improve development as it proceeds. Too often an end state is described and sought after, a set of operations intended to reach the end state is applied, and the whole system is set in motion to await evaluation at the end of the process. This "ballistic" approach does not permit sufficiently strong corrective measures to be taken during development, and can result in missed opportunities. A system of continuous monitoring permitting changes based on feedback information is a formative evaluation system. Further study of institutional structures that will increase the use of such evaluation should be encouraged and supported.

There is a significant problem associated with summative evaluation. When development is completed, no one appears interested in the evaluation, particularly the parties responsible for the original proposal and the development. If there are significant negative social or environmental consequences, preventive action is clearly out of the question; only costly corrective action can be taken. Frequently the cost is too much to bear, forcing populations to live with the results as they are.

Formative evaluation is not without problems either, but further study will serve to resolve or at least ameliorate them. Administrators responsible for decisions to alter a course once started often find the process too difficult to institute and therefore place less credence in information indicating the need for change. Sometimes the information may not be understood by the decision-maker, or he may not be in a position to implement the change directly; it may deal with a different time frame or involve larger systems than those in which the decision-maker acts. Sometimes not only must a course be changed, but the goal itself may be found wanting. Such a situation constitutes a serious threat for the individual and for any institution or governmental agency involved in development. Finally, formative evaluation holds the threat of "looking over the shoulder" of those engaged in action or development; this threat is not welcomed by most individuals and institutions.

Most of these problems are inherent not only in formative evaluation, but in evaluation in general, and they must be solved whatever evaluative method is used. The aspect of continual threat generated by formative evaluation is perhaps its most significant drawback and should receive the most attention in research on ways to introduce appropriate evaluative techniques for environmental change.

Before any development is proposed for acceptance, there are standards which must be met. Frequently in land development, broadly stated zoning and other codes are the only standards that contain or direct development. As often as not, such codes do not take into account the carrying capacities or limiting conditions of the environment; economic feasibility remains the only guide. Classic cases of overzoning were not in the least restrictive in New York City and Chicago; other factors prevented development under those conditions.

Means other than zoning codes should be studied to contain and direct development. The concept of the impact statement offers a promising replacement for code systems lacking flexibility and control. The application of impact statements has been criticized on several counts: inadequate evaluation of social impacts (Johnson, 1972); limited precision and structure of the statement (Robkin, 1972); and the general objection that diversity is not included as a variable. Further, enforcement of impact statements to date has been universally weak. Study of these and other aspects of the methods and rationale of impact statements will expedite their acceptance and improve their usefulness.

Impact statements presently in use and under development deal with rather diffuse problems in almost clinical situations. Some of the difficulty stemming from diffuseness may be resolved by research on models to be included in continuous monitoring and formative evaluation. Such technical studies must be accompanied by a data base for the general public. Those on whom development has impact should be allowed to make less formal but potentially very substantial contributions. "Grass roots" data sources can reduce the cost of data collection, despite the initially low quality of the data (Lewis, 1972). More importantly, increased awareness, improved skill and more direct action among the general public may result and make the establishment of an ecological conscience more probable.

SUMMARY

If we are to live in balance with our environment and create a total system capable of regenerating itself and evolving indefinitely, we need to be fully aware of the factors affecting: the formation of our values, the formulation of goals for the urban ecosystem, and the implementation of appropriate policies and programs. The basic goals of sustaining and enhancing human life are easily stated, but difficult to realize. We know that our material and energy resources are not inexhaustible and that their total depletion will create desperate situations; accordingly, the goal of sustaining human life requires that we begin to strive for a dynamic equilibrium in which our limited resources may sustain us indefinitely.

There is disagreement, certainly, about what constitutes "enhancement of human life", and about what programs will lead to its attainment. Human life, however, must be raised above the subsistence level if we are to survive as whole human beings.

Underlying any discussion of human goals, of course, is the question of human values. The way we choose to live is determined in large part by the values we hold; if our values are ecologically sound, our lifestyles should be also. Many of the values currently held, however, encourage lifestyles detrimental to the environment. Our apparent addiction to unbounded choice, for instance, is a key factor in the rate at which our resources are consumed. Likewise, the great emphasis placed on the economic value of open land in the city often blinds us to its ecological and socio-psychological values. We must begin to look at all facets of the urban environment in terms larger than the economic. This approach will require changes in our values.

But we cannot change our values at random. Ideally, an involved community educates itself on environmental and biological matters, sets ecologically sound goals and implements them. Many existing institutions, government and private, are unable to effectively facilitate citizen participation or to respond to the issues raised by those citizens and the requirements of the urban ecosystem itself. Often our institutions are strapped by jurisdictional problems. Responsibilities for facets of the ecosystem are fragmented among several narrowly conceived and narrowly defined institutions and institutional levels; the boundaries of the system seldom correspond with the boundaries of responsibility. There are few information input mechanisms furnishing adequate data about and interpretation of the state of the ecosystem to institutions and their decision-makers; there are few information output mechanisms translating the data and making them readily accessible to those affected by environmental decisions.

There have been successful attempts to create institutional structures which are responsive to the concerns of various segments of the urban population and which do have jurisdictional boundaries closely conforming to the ecosystem they are influencing. Likewise, there have been attempts to give urban ecosystem professionals a role in the acquisition of information and the formulation and implementation of urban policies. New skills are required to link various urban organizations; urban ecosystem professionals can make this task easier. Professionals can also provide and interpret needed information about the state of the system. That information, transmitted to the public by community information networks, can be used to draw attention to problems and issues which otherwise may be overlooked. Community design/development centers are also required; they can serve as the organizational link between various community organizations and can assist in the acquisition of needed skills and their application to high priority community projects. These centers can play significant roles in the formation of socially responsive and ecologically sound urban ecosystem policies.

If our efforts are to meet with success, however, we must: 1) develop ecological accounting systems which measure not only economic but also ecological, socio-psychological and other costs of maintaining the urban ecosystem; and 2) begin continuous monitoring of changes taking place in our environment and use that information to guide and control future developments. We require that the full range of tradeoffs related to a particular project be investigated and weighed in the decision-making process and reliable indicators of ecosystem health be developed. Further we require use of formative evaluation to assess particular projects rather than summative evaluation. Monitoring of developments should be continuous and indicated changes must be made when they are needed. We can no longer afford to wait until the final outcome of a project to assess it; the results may be unsatisfactory and corrective action often too costly.

The key facets of the urban ecosystem requiring continuous monitoring are its populations, its physical structure and function, and its water, energy and materials flows. With comprehensive knowledge of these components, we will be better able to formulate ecologically sound policies and to live symbiotically in the urban ecosystem.

RECOMMENDATIONS

Policy Recommendations

I-1. We recommend that institutions funding research projects give priority to proposals in the following categories:
a) Studies dealing with channels of communication for the immediate dissemination and subsequent use of on-the-shelf information.
b) Studies using appropriate monitoring and feedback techniques to insure ecologically sound results.
c) Studies seeking to expand the knowledge of new technology and new methods of teaching or communicating for use in the near future.
d) Studies examining the employment of such criteria as long life, low energy use and flexibility in various incentives for production or development, e.g. tax structures and pricing systems.

I-2. We recommend that urban institutions, agencies and professionals consider strategies which:
a) Concentrate control over policies affecting ecosystems at lower levels of government, or redesign large-scale units to be sensitive to local problems.
b) Encourage mechanisms by which governmental institutions may regulate factors affecting an urban area but originating outside its political boundaries.
c) Provide the public with current information about the state of the urban system and the implications of significant public and private projects.
d) Facilitate public participation in governmental decision-making and in evaluation of tradeoffs among conflicting values.
e) Support citizen organizations acting to maintain and enhance urban ecosystems.
f) Establish channels for the exchange of information from the neighborhood to the national level.
g) Utilize control mechanisms and decision-making processes that are visible, identifiable, accessible and politically responsive.

I-3. We recommend that urban open space be reclassified to differentiate between generative and regenerative land uses; and, further that the ecological and socio-psychological values of regenerative land functions (as well as the more usual economic values) be fully considered in planning.

I-4. We recommend that emphasis be given to planning which provides for alternative living patterns for individuals; planners can play a constructive role in creating options by facilitating community dialogue.

I-5. We recommend continuing and increased support for both formal and informal environmental education programs at all age levels.

I-6. We recommend that free-flowing community dialogue be recognized as a necessary precedent to formal decision-making by elected officials.

I-7. We recommend public support for urban development/design programs or centers within which urban professionals can perform their tasks.

I-8. We recommend that policy-makers recognize urban environmental and cultural diversity and consider the various perceptions of urban residents during decision-making.

Recommendations for Action

I-9. We recommend action to:
a) Establish ecological instead of economic standards for evaluating development proposals.
b) Replace energy-intensive activities with labor-intensive ones.
c) Encourage progress toward a steady-state economy.
d) Replace zoning codes with impact statements, performance criteria, and land-use analysis.
e) Expand public environmental data collection systems.

I-10. We recommend that formative evaluation (assessment intended to improve development as it proceeds) replace summative evaluation (simple comparisons of before and after conditions).

I-11. We recommend creation of mechanisms at the regional or local level to provide the public with adequate and unbiased ecological information and to detail the implications of proposed public and private projects.

I-12. We recommend development of techniques for participatory planning and of a network for community information exchange; these techniques may include: the use of interactive media, periodic issue balloting, interactive data bases and community cablevision.

I-13. We recommend establishment of "community places" with portable equipment to record community situations, or interactive computer graphics to display environmental indicators and to provide information on institutions, their programs, products and services;
we further recommend that the "community place" be equipped to record citizen opinions, complaints or suggestions.

I-14. We recommend immediate development of a national community information network; soon thereafter, a global network of information exchange should be established. The global information network should utilize and contribute to the Environmental Information Referral Service in Geneva, Switzerland.

I-15. We recommend the establishment of community design/development centers as useful institutional structures for linking formal and informal institutions already in existence and for mobilizing their resources and energies; such centers can and will play significant roles in the formation of socially responsive and ecologically sound urban ecosystem policies.

I-16. We recommend the establishment of

programs to give interdisciplinary and community-related training to urban professionals.

Research Recommendations

I-17. We recommend intensified research on:
 a) Measures of ecological efficiency.
 b) Maintenance costs of urban centers and urban regions.
 c) Criteria for a range of ecologically sound life styles.
 d) Minimum needs for social and physical data banks.
 e) Modeling at local levels to emphasize stability and predictability.

I-18. We recommend research to determine the basic design criteria for a community information network and community information nodes.

I-19. We recommend examination of the relationship between institutional responsibility and arrangements and the spatial and functional patterns in urban ecosystems.

I-20. We recommend research to substantiate the relationship between existing institutional structure and the failure of institutions to produce effective urban environmental policy.

REFERENCES

Anon. 1973. Agenda for the 70s. Center for Metropolitan Studies. Metropolitan Bulletin No. 10, March/April.

Barkley, Paul W., and David W. Seckler. 1972. Economic growth and environmental decay. Harcourt Brace Jovanovich Inc., New York.

Bauer, Catherine W. 1961. The urban octopus. Pages 117-122 *in* David Brower, ed. Wilderness: America's living heritage. The Sierra Club, San Francisco.

Berle, A. A. 1968. What GNP doesn't tell us. Saturday Review 51(31 August, 1968):10–12, 40.

Carson, Daniel H. 1974. Human habitat and environmental education: analogies and approaches. *In* J. Cairns, ed. The environment: conflict, costs and action. Decker Publ., New York. In press.

Carson, Daniel H. 1972. Residential descriptions and urban threats. Pages 154–168 *in* J. Wohlwill and D. Carson, eds. Environment and the social sciences: perspectives and applications. American Psychological Association, Washington, D.C.

Cheek, N. H. 1972. Variations in patterns of leisure behavior: an analysis of sociological aggregates. *In* William H. Burch, N. H. Cheek, and Lee Taylor, eds. Social behavior, natural resources and the environment. Harper and Row, New York.

Cooper, William E., and Raymond D. Vlasen. 1974. Ecological concepts and applications to planning. *In* D. M. McAllister, ed. Environment: a new focus for land-use planning. RANN National Science Foundation, Superintendent of Documents, Washington, D.C.

Downs, Anthony. 1972. Alternative urban development patterns. Paper presented at School of Architecture, University of Wisconsin-Milwaukee, March 26, 1972.

Fried, Marc, and Peggy Gleicher. 1962. Some sources of residential satisfaction in an urban slum. Journal of American Institute of Planners 27:305–315.

Granger, Charles H. 1964. Hierarchy of objectives. Harvard Business Review 42(3):63–74.

Hardin, Garrett. 1972. Exploring new ethics for survival: the voyage of the spaceship Beagle. Penguin Books, Inc., Baltimore.

Jacobs, Jane. 1961. The death and life of great American cities. Random House Vintage Books, New York.

Johnson, Per K. 1972. Social aspects of environmental impact. Pages 87–90 *in* Robert B. Ditton and Thomas I. Goodale, eds. Environmental impact analysis: philosophy and methods. University of Wisconsin, Green Bay, Wisconsin.

Johnson, Warren. 1971. The guaranteed income as an environmental measure. *In* W. Johnson and John Hardesty, eds. Economic growth vs. the environment. Wadsworth Publishing Co., Belmont, California.

Klausner, Samuel Z. 1970. Thinking social-scientifically about environmental quality. Annual of American Academy 389:1–10.

Lee, R. G. 1972. The social definition of outdoor recreational places. *In* William H. Burch, N. H. Cheek, and Lee Taylor, eds. Social behavior, natural resources and the environment. Harper and Row, New York.

Lewis, Philip R., Jr. 1972. Data needs and data manipulation. Working paper 8C. University of Wisconsin-Madison, Institute of Environmental Studies, Madison, Wisconsin.

Lipowski, Z. J. 1970. The conflict in Buridan's Ass or some dilemmas of affluence: the theory of attractive stimulus overload. American Journal of Psychiatry 127(3):273–279.

McAllister, D. M., ed. 1973. Environment: a new focus for land-use planning. RANN National Science Foundation, U.S. Superintendent of Documents, Washington, D.C.

Meltsner, Arnold J. 1972. Political feasibility and policy analysis. Publ. Admin. Rev. 32(6):859–867.

Newman, Oscar. 1973. Architectural design and crime prevention. U.S. Department of Justice, USGPO, Washington, D.C.

Robkin, Eugene. 1972. Some thoughts on impact statements and mathematics. Pages 91–93 *in* Robert B. Ditton and Thomas I. Goodale, eds. Environmental impact analysis: philosophy and methods. University of Wisconsin, Green Bay, Wisconsin.

Rokeach, Milton. 1968. A theory of organization and change within the value-attitude systems. Journal of Social Issues 24(1):13–33.

Rokeach, Milton. 1970. The measurement of values and value systems. *In* G. Abcarion, ed. Political participation and social-psychological processes. Charles E. Merrill, Columbus, Ohio.

Smith, Fred. 1972. Man and his environment: a manual of specific consideration for the seventies and beyond. Man and His Environment Project, Rockefeller Plaza, November, 1972, New York.

Smith, James N. 1972. The Gateways: parks for whom? Pages 213–216 *in* National parks for the future. The Conservation Foundation, Washington, D.C.

Sohn, Lewis B. 1972. Chairman 22nd report of the committee to study peace. The United Nation and the environment. The United Nations, New York.

Steinhart, John S. 1972. Institutions and the generation of purpose: whose environment gets managed and for what? Prometheus 1(4):1–27.

Stainbrook, Edward. 1968. Human needs and the natural environment. *In* Man and nature in the city. Bureau of Sport Fisheries and Wildlife, USGPO, Washington, D.C.

Strong, A. L. 1970. Crisis mentality and the deteriorating environment. *In* Roger Revelle and Hans Lansberg, eds. America's changing environment. Houghton-Mifflin Co., Boston.

Sudia, Theodore W. 1971. Man, nature, city: the urban ecosystem. National Park Service, USDI. Urban Ecology Series, No. 1. USGPO, Washington, D.C.

Tankel, S. B. 1963. The importance of open space in the urban pattern. Pages 57–72 *in* Lowdon Wingo, Jr., ed. Cities and space. Johns Hopkins University Press, Baltimore.

Terlecky, Nestor E. 1970. Measuring progress toward social goals: some possibilities at national and local levels. Management Science 16(12):765–778.

Weiss, Carol H. 1971. Organizational constraints on evaluative research. Columbia University Bureau of Applied Social Research, New York.

Wittenberger, Avery. 1973. Back to Eden: parks to let more grass grow wild. Milwaukee Journal, Accent Section, p. 1, June 7, 1973.

SECTION TWO:
Components:
Population, Physical Structure, and Resources

The chief components of the urban ecosystem are its populations, both human and non-human, its physical structure and its resource flows. If we understand the nature of these components, their interrelatedness and interdependence, and their limitations, we can better define ecologically sound courses of action.

Urban populations are perhaps the most difficult component to deal with. Actions intended to affect population size, for instance, have religious or moral dimensions. Actions meant to remove inequities resulting from race or socio-economic status, or to redistribute the human population more hererogeneously in the urban system, may meet cultural or economic resistance. These are factors operating within the urban population. Further, the city has undeniable psychological effects on its residents, effects we do not entirely understand. Moral and social involvements may be avoided. Courtesy may be forgone; strangers may be ignored. Urban residents are often unable or unwilling to take advantage of museums, theaters, parks, and colleges which could enhance life in the city.

In the central city, the pressures generated by high human density are frequently exacerbated by poverty and racial differences. In 1972, for instance, 70 per cent of all murders in Atlanta, Georgia, were committed in homes in the inner city; the killings stemmed from trivial issues. Mental health problems and alcoholism are significant problems.

The numbers, density and heterogenity of the urban population affect all aspects of city living. As the size of the population increases, people tend to compartmentalize their interests, behaviors and attitudes. Milgram (1970) terms such responses "adaptations to overload", noting that they are often characteristic of city dwellers. Most individuals, he says, avoid moral and social involvement except with their closest associates.

Studies of the effects of density on man have produced varied conclusions. Ehrlich and Freedman (1971) studied crowding in different-size rooms over a four-hour period. They found that groups of females adapted better to crowded conditions than groups of males. The latter became more competitive and less friendly. In neither sexual group did crowding seem to affect physiological responses. Ehrlich and Freedman concluded tentatively that density alone does not increase the incidence of human social pathology.

In contrast, a study by Galle, Gove and McPherson (1972) examined 75 community areas of Chicago and considered correlations among density, fertility, mortality, child care, asocial aggressive behavior and psychiatric disorders. Persons per room, rooms per housing unit, housing units per building and buildings per acre were the measures of density. The researchers found that the number of persons per room was the most important factor in determining cases of social pathology. The only exception was in the case of admissions to mental hospitals, where rooms per housing unit displayed the highest correlation. Admissions were also highly correlated with the percentage of persons living alone. These workers caution that such correlations do not assure cause and effect links but do indicate directions for future investigation into causal forces. The relationship between density and income level should also be studied. Such research has important implications for the re-evaluation and planning of health facilities and for preparing building codes.

Foa (1971) notes that social problems cannot be solved exclusively by the better distribution of economic resources. To some degree, man has adapted to urban living and may be able to adapt to an even more urbanized existence. There is considerable difference, however, between *may be able to* and *must* adapt. Empirical measurements of the city's effect on human behavior are urgently needed.

Human density studies, both demographic and behavioral, are limited. Although density effects are usually examined by those studying other animal populations, the density of human populations is often neglected. Human density affects many facets of urban ecosystems, particularly resource use and residuals discharge. The relationships between urban density and human behavior clearly demand increased study. Because urban planners, in particular, influence the density and size of city populations by influencing land use controls and taxation and transportation policies, research into the specific impact of such policies on density would be of significant use in city planning.

Large cities have lowered transportation costs by locating economic activities near each other. However, the ecological and economic costs of population size and density have often been overlooked. Several cities now exceed the carrying capacities of their readily available air and water. These cities have considered limiting economic growth to control population. It is essential that teams of hydrologists, meterologists, civil engineers, urban geographers, ecologists, and economists make detailed analyses of the present capabilities and the long-run requirements of our major cities and evaluate available alternatives.

As density increases, various negative feedback mechanisms, such as higher mortality rates, reduced immigration rates and increased emigration rates, may begin operating to reduce the population growth in a particular region. Mechanisms operating directly on birth rates may include economic factors, the cost of educating children, lack of space for housing, and loss of job opportunities. The nature and quantitative values of these feedbacks require study by teams of social and natural scientists. In many mammals, when population size increases, a physiological feedback mechanism becomes active to reduce birth rates and raise mortality rates, reducing the growth rate. There is little evidence as yet to postulate a similar mechanism in man. Thus we cannot assume that a negative physiological feedback mechanism will restrict the growth of human populations in cities. Further investigation is needed.

DETERMINANTS OF URBAN POPULATION SIZE

Many urban ecosystems are rapidly approaching their carrying capacities. They seem increasingly unable to adequately sustain human life and provide for the activities that man values. Pressures caused by natural increases in numbers and by high rates of migration contribute significantly to this situation. Research on methods and incentives to achieve zero population growth and to effect a drastic change in migration patterns should receive the highest priority.

The natural rate of population growth is a product of fertility and mortality rates. In a confined area such as a city, these rates would be the primary determinants of population

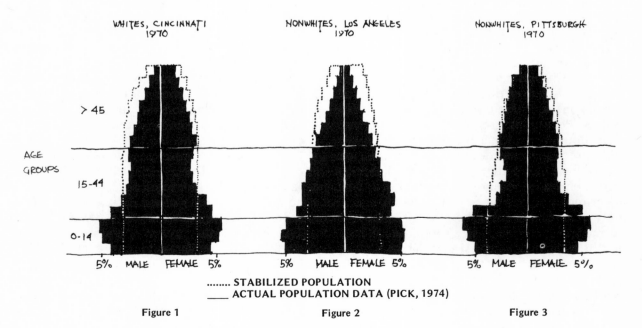

.... STABILIZED POPULATION
____ ACTUAL POPULATION DATA (PICK, 1974)

Figure 1   Figure 2   Figure 3

density if the area were closed to migration. In some urban regions of North America, the fertility rate has now approached replacement values, or the number needed to maintain the size of the present population (see Fig. 1), and populations are nearly stabilized; increases arise only from immigration. In other areas, however, fertility rates in excess of replacement values continue to exacerbate population density problems (see Fig. 2).

One of the first indicators of excessively high population pressure is an unusually high level of unemployment. In some cases, unemployment may rise locally to 25 per cent of the labor force while the national average remains at 6 per cent. The high level of unemployment precipitates mobility and emigration greatly exceeds immigration for specific age groups. **Figure 3** illustrates this condition in Pittsburgh in 1970, showing the loss of males and females in the 25-44 age group. Creation of work opportunities alone will not alleviate the situation. It may induce, rather, a rapid reversal of net migration trends, more persons coming into the system than leaving it; the size of the population will expand even further, increasing the problems of density.

Unemployment also has less visible effects. For every person who succeeds in leaving an area of low opportunity, there may be others unable to leave, who will remain in a stress situation. Correlations between population density and antisocial behavior have been documented (Fig. 4); crime places additional stress on those who cannot escape their urban plight and worsens an already bad situation. Reduction in the numbers of people living in parts of our urban systems might enhance the quality of life, and might even be necessary to sustain life.

Migration is another major factor affecting the size and density of our cities. The populations of many urban areas have reached a steady state in terms of fertility, but migration continues to drive the numbers up (see Fig. 5). Some areas, like New England, have a net emigration, the net difference falling below natural increases. Some central urban areas in that region, in fact, have declining populations. Other regions, such as the Pacific Coastal region, have considerable immigration

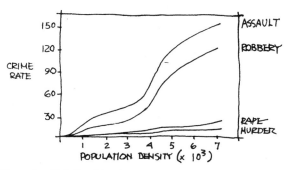

Figure 4 Relationships between density and antisocial behavior, Honolulu, 1970 (Boughey, et al., in press).

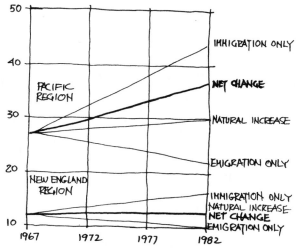

Figure 5 The influence of migration (Boughey, et al. 1973).

and the net growth far exceeds natural increases. Problems may arise from decreasing densities, but the most critical problems result from increased population densities. In such areas, human life becomes increasingly difficult to support and carrying capacities for such limiting factors as water and air are approached too rapidly. In these cities, technology may be inadequate even in the short run and the system may fail as a result of exploitation.

Relatively permanent long-range migrations are symptomatic of fluctuations in opportunities and amenities between different urban areas. To net reduce migration rates, these factors must be equalized. Short daily migrations (commuting to work or leisure activities) within urban areas generate high energy and material flows; to conserve resources, opportunities and amenities within urban areas must be equalized. Clearly such efforts are aspects of enhancing human life. To continue to encourage large and rapid migrations is to court environmental collapse resulting from excessive rates of energy and material use. To reduce migration without equalizing opportunities and amenities within and between urban areas, on the other hand, would place large numbers of people under mobility stress; they would live under poor conditions they could not escape.

Within the constraints of local limiting factors, urban ecosystems can become physically more integrated, utilizing their resources and energy more efficiently than at present. They can parallel natural ecosystems which (on the systems scale) are rather homogeneous within their constraints. Differences between urban ecosystems will begin to derive more directly from differences in their respective local conditions. Urban designers who now labor to construct differences—large skyscrapers, planned communities, industrial parks, massive freeway systems—will instead have to create differences at a more human level. Some planned urban differences burden local conditions, creating excessive ecological and social costs. Urban homogeneity (in the sense of a good blend of components on a neighborhood scale) may introduce some economic inefficiency. High rates of economic growth clearly cannot be sustained and at the same time permit a high level of environmental quality.

Births

Of the three most important factors in urban population size—birth, death, and migration—birth rates may be the most difficult to project into the future. Changes in fertility patterns are a result of various social and economic factors—some cyclical and others which are volatile and change readily. Mortality rates, on the other hand, are predictable, depending primarily on gradual developments in medical research, medical therapy, disease control and national health care. Migration, though it varies in volatility, seems more predictable than births in regions as large as cities and states, due perhaps to the presence of stable migration streams.

Many attempts have been made to account for fertility changes. Changing social attitudes, in part, contribute to the overall reduction in fertility. Fewer births result in changes in family size which, in turn, influence the needs of families for various-size homes and various kinds of community services. For example, Bumpass and Westoff (1970), who have examined the implications of providing reliable birth control, point out that people who are poor often have too many children too soon, i.e. before the family unit becomes economically stable. Their data also indicate that, in recent years, nearly 20 percent of all births have been unwanted.

Improved contraceptive practices and information will no doubt lead to a sharp reduction in the birth rate, a conclusion borne out by several studies (Jaffe, 1971; Clarkson et al., 1970) and by recent population estimates. In 1965, the "desired" family size was 3.4 children; a later study shows 2.5 children as the desired size. Though there are significant differences in the methods of the two studies, both note that "the number of children young couples desire before the experience of child-rearing may be largely a reflection of the observed size of the families of other couples with a similar social status." Cities designed to separate the poor will, it seems, perpetuate the development of larger families which will, in turn, promote continued poverty.

Biological factors also help determine family size. Lieberman (1970) has shown that the health and well-being of children and their mothers decrease with increasing family size.

Economic factors are considered important as well, although no positive correlation between the fertility rate and the state of the national economy has been demonstrated. However, in the latter half of the 1920s, prosperity was accompanied by a drastic reduction of the birth rate.

Considering both social and economic factors, Easterlin (1961) has developed some correlations based on the apparent U.S. 15-20 year cycles of increasing and decreasing fertility rates which began about 1855. The most recent cycle, starting around 1935, seems similar to earlier ones, although its span is considerably longer. Easterlin found the 20th century pattern while examining the economic and social causes of births to foreign-born urban, native-born urban and native-born rural women.

For many decades the national birth rate declined until it reached a low in the 1930s. Then it rose to a post-war high in 1957 and has now declined to a level (1973) nearly equal to that required to keep the population size static. This replacement rate in the U.S. is 2.11 births per woman per lifetime. If this rate continues and mortality remains constant, with no immigration, the U.S. will have a stable population after about 70 years. There would be, however, a substantial increase in total numbers during the interim, due to the large numbers now reaching reproductive age.

From Easterlin's work one could argue that we are now at the low point in the present cycle, and that in the future fertility will increase. Davis (1972) has suggested a possible lag due to the economic recession of 1969-70 which resulted in births being postponed, only to be made up later. The fertility effects of the considerable increase in the divorce rate are largely unknown, but possibly encouraged the recent lag.

The current quandary over the future direction of U.S. birth rates points to the inadequacy of fertility projection techniques. Keyfitz (1971) reviews available fertility forecasting techniques and concludes that the most appropriate model for post-war fertility might be the cyclical fluctuation of the fertility rate around the replacement level. He states further, though, that the amplitude and frequency of these waves are not currently known.

Nonetheless, accurate projection of fertility rates into the future would be useful to local, regional and federal officials in assessing the long-term total population and resultant economic and resource demands.

Deaths

Longer life and a greater chance for survival at any age are characteristic of people in the urban industrial society. We have greatly increased the survival rate of older persons, having raised the average age of death nearly as high as is possible for those who reach maturity. The greatest strides in public health, therefore, are to be made in the survival of the very young in our society. In primitive and even in agrarian communities, it was the losses in this pre-reproductive category that accounted for most of the deaths.

The theoretical bases for survival curves, modeling the age at which individuals in the population die, can be suggested from their shape and from our understanding of demographic concepts. In primitive societies, the mean life expectancy increases with age as the individual learns more, develops more resistance to disease and gains in strength and ability. In agrarian societies, on the other hand, the mean life expectancy remains constant; the probability of living one more year is as high at age 20 as it is at age 40. Thus, the major difference between the primitive and the agrarian society is the higher rate of survival among the young in the latter, due largely to a lower rate of disease.

The U.S. survival curves show a constant number of deaths in successive years, regardless of the increase in total population size. This constancy suggests the absence of effective feedback mechanisms relating total population to numbers of deaths. Each new individual contributes positively to the population rate.

However, in our urban society, many of the young still die from genetic deficiencies and childhood cancer; we now control infectious diseases like smallpox, scarlet fever and polio. The diseases of aging and the accumulated effects of accidents and exposure to urban conditions also become important mortality factors. Ford (1970) has called for increased research to identify the changing sources of urban mortality and to point out the needs of both older and injured people as they relate to other societal needs and values.

Schwirian and Lagreca (1971) have assessed the relative effects of demographic, environmental and social factors on death rates in urban areas. In looking at various geographical sub-areas, they found that housing and age had a direct influence on death rates, while economic factors seemed to operate through the housing variable. Additionally, they found the effects of housing situations and economic status were accentuated in those areas with comparatively young populations.

When we compare the fertility and survival curves of an industrialized urban population, we see that additional individuals do not necessarily provide added economic security to the population. This and the longer periods of dependency of some segments of the population (children up to 18 years and adults over 65) indicate that in present patterns a smaller and smaller number of individuals is supporting a larger and larger portion of the total population.

Migration

Of the factors influencing the urban population, migration is often the most important. Migration in the United States has both international and regional components. In the early history of the U.S., international migration was highly significant, as it is today in Argentina and Israel. In recent decades, though, the international flow into the U.S. has been so slight compared to the size of the total population (approximately .001–.002 of the total population) that it is unimportant.

Studies of migration between regions in this country have examined demographic and social variables such as time, physical distance, age groups, ethnic groups, occupation, education, marital status, service in the armed forces, income, economics of regions, duration of residence and others.

Age is possibly the most important of the

variables. Consistent evidence indicates that the ages 20–32 and 0–4 (their children) show the highest migration rates (either between or within states); the lowest migration rates occur in the age group 50 and older (Bogue, 1969; Louder, 1971). But the peak of movement in these age groups reveals nothing about whether net movement is into urban areas, out of them, or balanced. The standard age-grouping for some categories—college students or construction workers, for example—must be modified for small areas governed by special situations—a university town or an area dominated by a federal project (Louder, 1971).

Another significant factor in migration is duration of residence. Recent studies (Morrison, 1969, 1971) have shown that a small segment of the population is made up of repeated movers, and that, conversely, increasing length of stay in an area tends to reduce the likelihood an individual will move. In a study based on Social Security data, for instance, Morrison (1971) found that the proportion of migrating males dropped from .404 for those with less than one year's previous residence down to .126 after four years' previous residence, and finally to .042 after eight years' previous residence. Such duration dependence has been included in a few models of geographic mobility (Boughey, Pick, and Schick, 1973); it has often been included in the more numerous models of social class mobility (McGinnis, 1968).

Ethnic grouping is another important factor in migration. The traditional movement of blacks from the southern to the northeastern, north central and western states continued into the 1960s (Beale, 1971; Hauser, 1971). However, different patterns of movement can be anticipated as the proportion of blacks in the South continues to drop (77 per cent in 1940 to 59.9 per cent in 1960 to 53.2 per cent in 1970), and as increasing numbers of black persons in northern and western cities form new migrant streams.

One of the most detailed studies of smaller scale ethnic movements in a metropolitan area was done for Los Angeles County (Von Arsdol and Schuerman, 1971). The city was classified into sub-areas according to age and divided into regional cores whose populations could be traced through three decades. In newer areas there were indications of some increased mixing of ethnic groups. But the study revealed a continued overall picture of urban segregation.

In interstate migration, the variables of occupation, education and service in the armed forces form largely expected patterns. The yearly migratory rates for four broad job categories are as follows: white collar, 4.1/100; manual, 2.7/100; service, 1.8/100; and farm, 1.1/100. In the subcategories of white collar workers, we find the highest movement among professionals and technical workers (5.2/100). This movement of the highly skilled has been studied by Packard (1972). He notes the emergence of several categories of people who migrate, among them middle-level management personnel from large companies and skilled military personnel. An expanded analysis of such emerging patterns would aid communities in determining their labor supply.

Education is an evident factor. The migration rates for college-educated people (one or more years) is four times as great as for those who have only an elementary level education. Increasing educational opportunities in urban areas imply further increases in mobility among urban dwellers.

Over the last fifteen years, several models have been constructed to account for interregional movement in the U.S. Stouffer (1960) devised a two-region model consisting of a stream of migrants between two cities that was inversely proportional to the number of competing migrants and the intervening opportunities. More recently, many two-region formulas have been constructed to predict transfer of people. A recent regression analysis of movement between two Standard Metropolitan Statistical Areas (SMSAs) (Greenwood and Sweetwood, 1972) considered these variables: number of individuals in the two areas; highway distance between the areas; median income of residents; and per capita local government expenditures, minus welfare. Evidence indicated that migration was: 1) deterred by distance; 2) from low to high income localities; 3) to areas with high per capita government expenditures; and 4) to areas with temperate climates.

Despite their valuable contributions, Stouffer and the authors of these succeeding studies ignored one of the most important migration determinants, that of age. However Rogers (1968), using a matrix model of population transfer between many regions took into account age-specific immigration and outmigration, and birth and death rates. In his projection, beginning rates were used for the entire period; the in- and outmigrants were not balanced. Tabah (1968), using Mexican data, included male/female, employed/unemployed, and urban/rural considerations. However, his projection matrix became too large for manual calculations and computer processing of the data apparently was not attempted.

A recent computer model of migration between nine regions in the U.S. (Boughey, Pick and Schick, 1973) took into account age, sex-specific in-migration and outmigration, and other processes to project regional population over the next fifteen years. One closed version of the model balanced the whole system after each five-year projection period, so that outmigrants from all regions equalled immigrants to all regions. An open version based immigration on the total size of the receiving region and followed immigrant groups after arrival; this allowed the researchers to consider highly mobile, chronic movers as well as normal immigrants.

To date, systems constructed for modeling migration have been limited in scope. A broad, interdisciplinary model of movement, including such aspects of the urban ecosystem as economics, politics, climate, geography, and population density and distribution, would increase the relevance and accuracy of inter-regional projections. The U.S. Census at present has limited value for collecting facts on transfer of population; the broad model, necessarily constructed on a computer, would probably be untestable because of the restricted data base the census provides.

The importance of migration to cities lies in the role that waves of immigrants play in forming young cities and in molding the ethnic and occupational distribution patterns of more mature cities. Migration also affects the location of population in relation to resources and economic production, and the relationship between cities and the nation's diminishing rural areas. Understanding migration is essential to understanding urban population processes.

DISTRIBUTION TRENDS

More than 70 per cent of the people of the United States live within the boundaries of Standard Metropolitan Statistical Areas. Even those living outside SMSA boundaries frequently work within and receive much of their material and information from the metropolitan area. The urban resident has, on the average, less acreage and more capital at his direct disposal than does the rural resident.

Between 1960 and 1970, more than 80 per cent of our national growth took place in SMSAs, and these areas increased in number from 212 to 243. Much of the growth was in "suburbia" where more than 76 million people now live. Large central cities tended to fluctuate, and now contain 64 million people. Migration from non-metropolitan areas accounted for one fourth of the gain in metropolitan areas; the excess number of births over deaths accounted for the remaining three-fourths.

Populations of the rural areas within SMSA decreased during the 1960s. Taeuber's research (1972) indicates also a general migration toward the coasts, most of it to the large cities found in those regions; it is difficult to discern if the trend is coastal or urban.

Estimating the future distribution of our nation's population depends on realistic projections of total numbers. If replacement fertility (net reproduction = 1.0) continues to the end of this century, our population could still increase by 80 million, or approximately 40 per cent more than at present (Keyfitz, 1971).

This large addition to our population might be distributed in several ways. In the absence of or perhaps despite any national government policy, the population is likely to locate in present urban centers. Pickhard (1971) suggests that concentration in coastal urban centers is most likely.

The future population increment might be distributed to improve the urban-rural balance and population density patterns. Whether future distribution will be achieved consciously through government policies, like the fostering of "new town" programs, or unconscious-

ly through the economic and social situations which have governed distribution in the past (Ward, 1972), is uncertain at present. Whether planned or haphazard, however, the growth of urban populations will create differing urban environments, and could help either to resolve or to exacerbate present urban problems.

The problems of urban housing are closely tied to those of urban population growth. Most residential growth is taking place not in the urban core but at the city's fringes and beyond, helping supply housing for the poor and disadvantaged through a "trickle down" process. Families from lower income levels improve their situation by moving into those areas vacated by higher income level families moving spatially outward and socially upward. This process depends, however, on continued urban growth and expansion; unless downtown areas can be rejuvenated and made more attractive to higher income groups, the pattern will continue.

Slowing suburban expansion may reduce the housing available to the poor. Increased housing subsidies for low income families may be required to put them on equal footing with middle and high income families who presently enjoy an estimated $5.7 billion in indirect housing subsidies (in the form of deductible property taxes and interest payments).

Residents of inner city neighborhoods also lack adequate recreational space. The National Park Service is presently attempting to preserve or to clear and improve some urban land with its Gateway proposals (J. Smith, 1972) and the resulting recreational space will be nearer the inner city. Until now, the National Park System has been available chiefly to middle and upper income groups. Though this new project needs more effort and direction, it can serve as a base in providing needed recreational space to inner city residents.

Another experiment in urban living has been the development of "new towns". New towns can alleviate urban ecosystem problems only to a limited degree. They can accomodate only a very small percentage of the population—less than 4 per cent of the population by the year 2000 (Series E) and only about 20 per cent of the total population increase between now and then. New towns cannot entirely relieve population pressures in central cities and in some ways may worsen the situation. They must be heavily subsidized and they may draw attention away from urgent problems in central cities. Further, new towns are built most often for middle and upper income classes with little accommodation for the poor, further segregating these groups, their activities and value systems.

New town policies usually attempt to: 1) integrate the best city planning concepts; 2) provide a range of employment opportunities; 3) provide housing at a scale and price which allow for a wide diversity of people; 4) incorporate good public transportation and separated rights-of-way for pedestrian and vehicular traffic; 5) use new or experimental technology whenever possible; and 6) plan comprehensively before construction begins. Only the last of these policies may not be applied specifically to already existing cities.

Certainly new town planning and construction can parallel inner city planning. It should not, however, be emphasized at the expense of the inner city. New towns, growth centers, ocean cities, and other experimental cities should be considered primarily experiments, not panaceas for urban problems. Within our urban areas there is much space to be developed and redeveloped for generative or regenerative functions in the next few decades regardless of new town policies and progress. We sorely need policies directed toward creative growth to enhance the lives of urban residents.

Age Distribution

In the United States, the age distribution of the population up to age 50 is becoming more the product of past fertility changes than of international migration. On one hand, the expected proportion surviving to age 50 is high—86 per cent for males and 92 per cent for females in 1967 (Keyfitz, 1971); on the other, international migration has been low recently in proportion to the total population.

We can detect fertility fluctuations over the past 60 years or so by looking at 1970 age distribution patterns. There is, for example, a "depression age gap" caused by lowered fertility during the late 1920s and early 1930s. A pronounced reduction in births over the last ten years is also noted. Coale (1972) has done a detailed analysis of age distribution by changes in fertility and mortality curves.

Distinct age classes exist in the urban population and they are often segregated from each other. Before 1900, children born of a single marriage were often spread over several decades so that many households included three generations and some spanned four. Interactions between individuals of all age groups were common in those households, which remained intact for many years. The major provider in the family could change several times during the life of an individual. This is clearly no longer the case.

Since today's average urban parents are younger than those in the past, they produce more normal and healthier children. These parents are better educated and they tend to demand more medical and educational resources for their children. These resources are close at hand in the city and are available to more pre-reproductive age individuals. As a result, the young consume a large portion of the economic resources of society, particularly for education.

Today's post-reproductive individual can expect to live in the company of a larger proportion of his age group than was previously the case. Formerly the grandparents lived within the family unit; now these older people generally live in extra-familial situations, a condition unique to recent times and most common in America. They also consume a high fraction of certain services such as medical care.

Because of the special needs of older persons, we have built such facilities as retirement communities, rest homes and apartments for the elderly. Much housing today is built for young couples, particularly those with children; after the children leave and the parents grow older, the dwelling unit becomes a problem. It is frequently too large, or is not designed to accommodate the "invalid", lacking ramps or doors wide enough to allow for a wheel chair.

Clearly, urban society is faced with an increasing proportion of its population falling into the post- and pre-reproductive age-groups; further, an increasingly large proportion of "non-productive" individuals are now being supported by the decreasing number of productive ones.

Ethnic and Socio-Economic Distribution

Since the early settlement of our country, American society has been characterized by an isolation of classes—the poor, the rich, the middle class, the religious group, the ethnic group. Each has remained very distinct from the others.

Today nearly 12 per cent of the U.S. population is black, 5 per cent is Spanish-speaking, and slightly less than 1 per cent are oriental, Indian, or other ethnic group. In cities, these groups are frequently identified with particular neighborhoods which develop characteristic flavor and style. "Chinatown" may seem a unique portion of many cities; however, most segregation affects blacks and Spanish-speaking Cubans, Puerto Ricans and Mexican-Americans.

In suburban areas of relatively new western cities, the superficial mixing of various European groups is somewhat countered by the social segregation of church, business activity and social groupings. Cultural and genetic separation is often maintained despite the physical integration of housing and jobs.

The black population has not only become urbanized, but has also concentrated in the central cities, increasing tenfold over the last 30 years, from 1.3 million in 1940 to 13.1 million in 1970. Blacks now account generally for 21 per cent of central city populations, and for 28 per cent of the population of larger cities (SMSA greater than two million). The majority of the populations of Atlanta, Gary, Newark, and Washington, D.C., are black; nine other large cities are more than 40 per cent black. While growth of black populations in suburbia has matched that of the whites, migration of blacks to SMSAs was twice that of whites. Almost all the blacks went to the inner city.

In El Paso, Laredo and San Antonio, Mexican-Americans constitute approximately 40 per cent of the population. They also make up 20 per cent of Pueblo and Albuquerque and nearly 10 per cent of most large cities in the Southwest (Galarza, Gallegos and Samora, 1969).

In many major cities, there are sizable ethnic groups. Many represent distinct socio-economic groupings having a distinctive income, religion, social structure and set of needs.

We have discussed ethnic groups at some length because they are easily identified and their characteristics are highly correlated with various social and economic factors. A large percentage of the population in the central city receives a poverty level income. Since migration is predominantly into the central city, we must begin to know more about the potential for upward mobility. Many of the nation's poor were poor in rural areas and they remain poor after moving into the city.

Galbraith (1964) suggests that even from a selfish point of view, it might be as wise to provide the poor with a decent life as to provide the upper class with a still better life. Ways to replace the economic burdens of maintaining central city welfare programs with the means for the poor to provide for themselves need further analysis.

Majority attitudes toward racial and economic minorities in a neighborhood play a significant role in determining movement within the urban/suburban complex. The inability of majority populations to cope with any substantial minority residence in a recurrent theme in urban population shifts; it reaches far back into American urban history. The study of these attitudes and the motivations of those moving to and from urban centers would prove useful in predicting the future migratory patterns of ethnic and economic groups.

Other Patterns of Distribution

Several other factors affecting the urban population need to be considered by planners. Some people live in the commercial and industrial sections of cities; others reside in suburban housing developments. Age, ethnic, socio-economic or religious segregation produces considerable variation in birth and death rates in various sections of metropolitan areas. An obvious example is the housing development limited to senior citizens or adults only; this population may have little need for public schools, but considerable need for medical care. If financially secure, the post-reproductive couple, for example, may be interested in theaters, high-quality restaurants and spectator activities; the young couple, on the other hand, may be more interested in schools and playgrounds.

In cities where people are some distance removed from their places of work, some children can live and grow up in pleasant suburban surroundings. Side effects, however, include large transportation costs in moving workers to and from their jobs, higher accident rates, poorer health conditions for workers in heavily industrialized zones, and travel stress. Separation from the working parent or parents may also result in psychological stress for the family remaining at home.

If industrial and residential patterns assure easy commuting, density problems may result. For example, Eisenbud (1970) claims that one-third of all New York City's particulate emissions in winter come from residential space heating. He states also that "the main source of carbon monoxide exposure for city dwellers is apt to be cigarettes." Clearly, these non-industrial, non-automobile related phenomena deserve additional consideration in the design of air movement in buildings.

Doxiadis (1968, 1970) reviewed the development of human settlements and suggested that the pattern of population distribution is related to the need and ability of the individual to transport himself. Historically, most cities developed with a radius which could be covered in a ten minute walk; between cities the usual walking time was no more than one day. Accordingly, cities had a limited number of inhabitants, about 50,000. Capitol cities, however, often became much larger; in such cities, long straight throughways were built to allow rapid movement of animal-drawn vehicles. Mechanized transportation followed later.

The capability for rapid movement of large numbers of people should be an objective of urban planning. Too often, however, we plan piecemeal, ignoring total population size and the relationships of various sections of the city to each other. Doxiadis maintains that services are frequently designed for certain classes alone; he calls for real freedom of choice for everyone. An optimum urban size, he states, is not as critical as are healthy and prosperous living conditions for urban inhabitants.

Dividing the city into a series of villages, in turn divided into regions and neighborhoods, may resolve some urban problems by regulating patterns of distribution. F. Smith (1972) suggests an optimum average density of 15 persons and 5.6 living units per acre, 500 persons per neighborhood, 2,875 persons per region, and 11,500 per village. He suggests that "new towns" of this type can serve as experiments or "pilot plants" to provide new perspectives on the problems of existing cities. Further examination of this kind of urban organization may help planners, businessmen and city service officials.

Clearly, problems of population density and distribution are not easily solved; much essential information is lacking. We need to be able to gauge birth rates and migration patterns, for instance, if we are to plan adequately for urban service institutions and transportation systems. Fertility, mortality, and migration remain key factors in urban population dynamics. Human population pressures affect the functioning of the entire urban system. We must begin to understand these pressures and to act wisely to relieve them.

NON-HUMAN POPULATIONS IN THE CITY

Consideration of urban population processes cannot be confined to the human element only. A substantial number of other organisms—plants, mammals, birds, reptiles and insects, etc.—are also found in cities. Many live outside man's direct control, as wild or feral species; others exist as pets or plantings. The increasing demand for pets and plantings suggests a deeply felt need of city residents to surround themselves with other living things; the need appears so great that it may be deemed essential. Despite this, study of urban non-human populations has barely begun.

The most familiar urban animals are pets. About 46 per cent of all U.S. households own a dog or other such pet (Feldmann, 1974) and 20 per cent (one in 14 persons) own at least one cat (O'Keef, 1972). The urban fauna also includes an unquantified but significant population of stray and feral cats.

Rats, mice, racoons, opossums, skunks, coyotes, bats, foxes, squirrels, rabbits and shrews are also abundant in the city. The numbers and distribution of these animals are determined by a neighborhood's age and degree of development. Larger older cities may have more rats than people.

We have no accurate count of house plants, ornamentals, shrubs and trees in the urban environment, but the annual cash expenditure for horticultural goods and services is at least four billion dollars. Untended plants, both native and escaped, also contribute significantly to the flora of the city, but their abundance and ecology are poorly known (Schmid, 1973). Many exotic species now widely distributed have become troublesome weeds.

A complete list of urban birds in America may include 200 species. Various species occupy their own distinct and specific urban habitats—paved commercial areas, parks, industrial sites, residential areas and waste disposal sites.

Mosquitoes, ants, termites, roaches, bed bugs, flies, fleas and lice are some urban arthropods influencing human health, economics, and social well-being. Likewise, bee, wasp, and scorpion stings contribute to health costs.

Dog ownership has increased 36 per cent in the last ten years presumably because dogs serve a deterrent to crime and as companions. There are an estimated 40 million owned dogs in the U.S. (Feldmann, 1974), approximately one for every seven humans (Beck, 1973; Marx and Furcolow, 1969). The urban environment is a hostile one for free-ranging dogs. In 1970-71 at least 58,000 dogs were destroyed in New York, 21,000 in St. Louis, and 13,000 in Baltimore. This rapid turnover reduces the average age of urban dogs, allowing an increased incidence of diseases transmittable to humans (Webb, 1965).

Dogs are gregarious, with about half of all free dogs roaming in the company of other dogs, facilitating the transmission of diseases from dog to dog. Gregariousness also fosters destructive behavior such as biting or disruption of trash. Dogs permitted to run free often knock over or open garbage cans; exposed trash becomes the primary breeding area for houseflies and a major source of food for rats (Beck, 1973). Loose dogs also pose serious traffic hazards. In Great Britain, dogs cause 6 per cent of all automobile accidents (Carding, 1969).

Disposal of waste is a serious problem with urban dogs. An average-sized animal produces 227 g (0.5 lbs) of feces and 720 ml (0.19 gallon) of urine a day. The fecal contamination of the city surface totals 125 tons/day in New York City, 25 tons/day in Baltimore, and 21 tons/day in St. Louis, and poses a problem in the handling of storm water runoff. Equivalent human contamination would not be tolerated.

Dog feces are also a major source of houseflies in the city (Quarterman et al., 1949) and a potential source of Samonella (Wolff et al., 1948) and Shigella (CDC office, personal communication).

The substantial benefits resulting from interactions between man and dogs might be achieved with a substantially lower dog population. Improvements in dog-licensing systems, regulation, and unowned dog control programs are necessary. The costs of insuring adequate dog control could be recouped in reduced disposal costs and reduced demands on public health facilities. Other pet species might provide similar benefits at lower cost.

Animal bite is a major source of injury in the city. The true frequency in the United States is not known; at best only about half the actual bites are reported. In 1971 nearly 40,000 dog bites were reported in New York City, while 7,000 were reported in Baltimore and 3,000 in St. Louis. This bite rate averages out to more than 50 per 10,000 people. These same cities, which all have serious rat problems, reported rat bite rates of less than 1 per 10,000.

Dog bites alone account for 1.6 per cent of all hospital emergency room admissions (Brobst et al., 1959; Beck, personal communication). Dog biting is not random. Children are the most frequent victims (Berzon et al., 1972; Parrish, 1959). In St. Louis the age-specific rate for children ages five through nine is 202 per 10,000; 2 per cent of the children in this age group are bitten each year (Beck, in progress). Among urban adult humans, bites are usually occupationally related—victims are letter carriers, utility meter readers, and service personnel. About 80 per cent of all reported dog bites involve laceration and contusions and about 10 per cent require surgery. Two percent are serious enough to require hospitalization or extensive surgery.

Disease is the chief threat posed to man by animals in the urban environment. Zoonoses are diseases transmitted to man from other animals; many animal parasites can cause human disease as well. The classical zoonoses related to urban animals, like plague and typhus, have been controlled in the United States. But other health hazards remain.

The fear and potential danger of rabies, for instance, adds greatly to the trauma and expense of any dog bite injury (Cereghino, 1970; Schunrrenberger, et al., 1969). Though rabies is now more commonly found in wildlife, especially in skunks and foxes, than in dogs, bites from wild animals account for only a small fraction of the reported cases. Dogs readily contract rabies from wildlife, and are most often responsible for a person requiring rabies treatment (Martin et al., 1969).

Dogs also harbor and transmit other diseases, among them tuberculosis, leptospirosisone and histoplasmosis. Since dogs may not show the symptoms, they are especially dangerous, particularly to children with whom they are often in close contact. (Hawthorne, et al., 1957; Hawthorne and Lauder, 1962; Hull, 1963.) In general, the magnitude of the health hazard posed by pet animals probably far exceeds that from independent animal populations.

Bird zoonoses appear minor when compared with the mammalian ones. The body temperature of birds is several degrees higher than in man, creating an effective impediment to disease transmission. However, several widely adaptable fungi infect birds as well as man and other urban mammals (Emmons, 1954; Emmons, 1955). Approximately 30 human fatalities per year are related to psittacosis, a virus infection usually resulting from contact with pet birds (Meyer, 1959). Likewise, birds play a vital role in the transmission of encephalitis. The wide mobility of urban birds may be a significant element in the epidemiology of this disease. Even so, the overall avian contribution to human disease is small; health-directed bird control programs produce small benefits per unit cost.

Plants, while not as damaging as animal populations, contribute their own share of problems to the urban environment. Plant-induced allergies constitute sources of discomfort, illness, and death and compound the problems of air pollution. Plants may produce significant quantities of allergenic pollen for extended periods.

Horticulturalists, landscape architects, and health professionals working together can reduce the burden of pollen-induced allergy diseases. Horticultural and medical research should focus on identifying species with irritating or non-irritating pollen, or reduced amounts of pollen. Plants with no allergenic characteristics can compete with the highly allergenic species and will grow well on non-cultivated city land. However, past experience, well-documented in botanical literature, suggests extreme caution in importing new species; emphasis should be placed, rather, on modifying characteristics of native plants.

Cost and Benefits of Urban Plants and Animals

Despite the hazards posed by nature in the city, it brings important benefits—both psychological and economic. The purchase of pets, planting stock, wild bird feed, fertilizers and soil builders, and payment for lawn service and veterinary care, are direct monetary exchanges included as benefits in economic indices like the Gross National Product.

The 38 per cent of urban American families who own dogs spend four cents of each grocery dollar on dog food (Beck, 1973). Another study indicates that 2 per cent of all grocery sales are for pet food (O'Keef, 1972). The urban plant industry has a yearly gross income of four billion dollars (American Council of Nurserymen, personal communication, 1973).

Plants bring other less obvious benefits to the urban environment. Energy conservation can be accomplished through the use of plants as wind screens, evaporative coolers (Schmid, 1973) and insulators (Horsbrugh, 1972). The affects of air and noise pollution can be reduced by using urban plantings (Schmid, 1973).

Like plants, animals can benefit man. They may serve as indicators of environmental quality. Plague antibodies found in domestic dogs, for example, alerted health officials who stemmed an epidemic on an Indian reservation (Archibald and Kunitz, 1971). Similarly, urban and rural dogs, which show noticeable differences in pulmonary diseases, have been a useful measure in determining human susceptibility (Reif and Cohen, 1970) and wild pigeons have been used as indicators for air pollutants (Tansey and Roth, 1970).

Urban animals, especially pets, increase costs of municipal services such as trash collection, street cleaning, park maintenance and mail delivery. They also increase medical-related expenses. The medical cost created by dog bites alone is estimated at more than one billion dollars per year.

The pest control industry and city governments expend immense sums of money, largely misdirected, to control relatively innocuous pests, while the more damaging urban animal wildlife, such as dogs, cats and rats, are receiving relatively little attention. Present control methods are economically inefficient. The funds to control disease-carrying pests could be allocated to projects designed to reduce the actual burdens of plant- and animal-transmitted diseases in the urban system. When a disease is rare, funds spent to improve treatment and diagnosis may be more effective than those spent to control the source.

Clearing and mowing of vacant land areas are costly to both public and private sectors of the urban economy. Development of ecologically stable communities using suitable ground covers, shrubs and similar horticultural alternatives could eliminate mowing costs without reducing benefits.

Substantial costs are also incurred in the cleaning and increased maintenance costs for buildings fouled by urban bird populations. Reductions in these costs will more reasonably be found in architectural design than in population control. But both depend on accurate knowledge of birds in urban environments.

In spite of the costs, plants and animals contribute significantly to the aesthetic and cultural enrichment of life in the city (Iltis, 1972). Benefits of plant-human and animal-human interactions go beyond the improvement of appearances. Man's positive relationships with plants and animals help to increase the habitability of urban environments.

Inner city gardening projects have benefited local neighborhoods and residents and have resulted in the reduction of vandalism, cleaner streets, painted houses, conversion of debris-laden lots into gardens and playgrounds and improvement of social interaction. Lewis (1973) relates these results to those from horticultural therapy in mental and physical rehabilitation hospitals and suggests potential benefits for such programs in correctional institutions and homes for the elderly. A few programs using pet psychotherapy, where patients interact with animals as a first stage in their therapy, already exist.

It is not entirely clear how the presence of other forms of life ameliorates stress and contributes to the quality of human life, nor how the benefits of these interactions can be enhanced. Nevertheless, current knowledge should be used and some reasonable assumptions accepted immediately in urban planning. Policies should be adopted which will encourage constructive interactions among plant, animal and human populations in the urban system.

Studies of man's influence on habitat and on the feeding and control of most urban organisms are still largely wanting. Animal life in the city remains less well-understood than wildlife in other environments, yet accurate estimates of urban populations are basic to rational management. Food webs, energy pathways, predation and urban wildlife habitats do exist, but the manner in which space and nutritional requirements are met is unknown. The impact of these species on man has not been adequately studied. Likewise, cost-benefit analyses of animal control programs are rare, although most cities have some data base to guide their expenditures. Only scattered information is available on the contribution of urban vegetation to air and water pollution, or conversely, on the potential contribution of vegetation in pollution absorbtion.

Animals and plants are important facets of the urban ecosystem. Their characteristics and roles must be better understood if we are to form ecologically sound management policies.

PHYSICAL STRUCTURE OF THE URBAN ECOSYSTEM

Cities at first were formed for religious, military or governmental purposes. Proximity also provided advantages for some trades; accordingly, people banded together to perform their functions more fruitfully. With the commercial and, later, the industrial revolutions, widespread urbanization began; by the last of the 19th century the developed nations of the world were more urban than rural. Today, nations beginning to commercialize and industralize are repeating that pattern.

Trade and production efficiencies enabled urban populations to live far from the sources of many raw materials needed to support life.

Governmental and social organizations in the city have been adapted to better provide needed services. Many people have been attracted to the urban center because it presents social and economic opportunities not available elsewhere. The urban dweller is frequently able to get a better job, obtain desired services more easily and live more comfortably than his rural counterparts.

Historically, when transportation and communication technologies were less well developed, the urban society was necessary to achieve economic efficiency and social opportunity. Now, however, improvements in transportation, changes in technologies of production and distribution of goods, improvement in communications, government programs pressing for widespread and readily available health services, schools and recreational activities have made it less necessary to live in a city to enjoy an urban life style.

Our society is characteristically urban wherever one lives. Though the urban center as we know it may become obsolete, the problems associated with it will probably continue as long as there is an urban society. Therefore, the problems of the urban structure and function will remain.

The physical differences between cities are largely the result of three factors: 1) the time of their most rapid growth; 2) their physical site; and 3) their situation or strategic location.

Differences in housing and the relative density of population primarily reflect the kinds of transportation and shelter that were available when the city grew; they also reflect the availability of resources at that time and the preferences of the inhabitants. Older cities, for example, tend to have narrow streets, small blocks, and dense housing, reflecting the pre-automobile culture. Newer cities or new sections of old cities, with wider streets, shopping centers, etc., reflect the automobile culture. Most cities are a blend of the two.

The site of a city refers to the immediate physical properties of its location. Many American cities date from the pre-railroad era and were sited on navigable waters. Sites of cities vary greatly; they may be hilly or flat, well or poorly drained or dry, and they may have widely different macro- and micro-climates. Site conditions are reflected in the layout of transportation routes and in the locations of various activities giving each city its distinctive characteristics. Its site also provides a city with distinct sets of problems in adjusting to its immediate natural environment.

The strategic location or situation of a city determines its growth and future. Cincinnati, located on the great north bend of the Ohio River and accessible to Ohio farms, grew early because of its favorable location. Produce from Ohio farms was shipped through Cincinnati down the Ohio and Mississippi to markets in the South. Other cities began as trade centers for farming communities; still others were gateways to international trade.

But as technology, the economy, and urban society have changed, the meaning of a city's strategic situation has changed also. Dense networks of high volume transportation routes now enable most cities to receive and ship goods as far as needed without disadvantage.

Accordingly, as more people have financial security and leisure time, the attractiveness of a site, particularly in terms of climate, seems to play an influential role in guiding the growth of urban areas. In the United States, the cities and regions growing more rapidly are those in warm climates.

A city's history leaves a heavy imprint on its present nature. Some cities grew rapidly because of their strategic location, but then they lost that advantage and declined. Others maintained their advantage despite technological changes. Still others have grown only slightly since their original settlement. Despite these variations, however, all cities have grown using common technologies and, for the most part, common societal and governmental organizations.

Cities are geographical entities and, as such, they can be described in terms of certain properties: configuration, which is the general shape and pattern of the city; density, or population per unit of space; differentiation, or the way in which various sections of the city are used; and internal connections, or the systems of highways and other pathways linking various parts of the urban area.

Configuration

Configuration, the general shape and pattern of an urban area, is the product of the natural physical structure of the site and the man-made overlay. Numerous natural factors affect configuration, among them physiographic features like the slope of the land and drainage and water patterns, the geological substructure, and the microclimate of the area. Biological factors do not greatly affect configuration but they may have some effect, e.g., the inclusion of grazing commons in the layout of a town or, more recently, the inclusion of green belts and city parks.

Man-made patterns imposed on the natural geography help to shape the urban configuration. Several geometric designs are prevalent in these man-made patterns. Rectangular grid systems have long been used. Radial configurations, dating from the early 18th century, are found in a few cities. The township and range survey system, used especially in the Midwest and western United States, produced a distinctive rectangular grid pattern in cities, except in those older sections where early Spanish or French settlers left their distinctive geometric imprint.

In recent years, suburban developers have departed from rigid street patterns. They now use patterns more suited to the geographical site and are also experimenting with new geometric patterns.

In the city, the patterns of streetcar tracks, suburban rail lines, and expressways also leave their mark.

Urban planning sometimes affects urban configuration. At the end of the 19th century, there was great activity towards preserving urban green space; some cities, such as Chicago, owe their extensive forest lands and waterfront parks to such efforts. The radial street patterns of Washington and Paris are the result of urban planning. There are other instances in which conscious planning has affected urban configuration.

Many trends in urban configuration are the result of planning decisions made outside the city. The location of an interstate highway system link, an airport, or a major industry affects the configuration of an urban area. Such decisions are too often made, however, not in the context of the specific urban area they will affect, but in the context of the region they service, often to the detriment of the city. If the responsible urban decision-maker could control such decisions, he would be better able to shape the form and configuration of his city.

Decisions about where to locate basic urban activities are usually made on the basis of access versus the price of land. Any activity at one location draws from or feeds into activities at other locations; the costs incurred in transportation between these activities play an important role in urban development.

If certain sections of an urban area offer more opportunities, either in terms of complementary activities or decreased access costs, development there will be accelerated. For example, historically industries developed along rivers because the availability of water provided both a necessary resource and lower transportation costs.

Density

The second element in an urban center's physical structure is density. Density commonly refers to frequency per unit area, as in residents per square mile. However, the increased upward growth of cities, with high rise apartments and skyscrapers, suggests that density defined in terms of volume may be more appropriate. It may also be necessary to define density in a temporal framework, as when the design of a physical structure must accommodate temporal fluctuations in hour-to-hour densities. All structures have both upper and lower density limits; day-to-day and hour-to-hour "density smoothing", as provided by staggered work hours, may improve the efficiency of urban structures. Nevertheless, the trend presently seems to be increased, rather than decreased, temporal variation in densities.

The major human density patterns in urban areas are found by examining where people live, where they work, and where they take part in other activities such as recreation and shopping.

Using a fairly gross scale, on the order of square miles, urban residential densities can be described in a series of schematic graphs (Fig. 6).

These density graphs and census information describe one of the major trends in urban growth. The last two U.S. censuses have indicated that the size of inner city populations has remained relatively stable and that most population growth is occurring in suburban areas. As a result, current urban development is consuming land at an accelerating rate, a situation commonly referred to as suburban sprawl.

A counter trend over the past ten years, however, is the increase, by about one-third, of the number of multiple unit dwellings, while the number of individual residences has remained about the same. This trend is not reflected in the density curves of **Figure 6** which only approximate densities along radial directions. If densities are examined at specific locations in the suburban belt, peaks in residential populations become evident.

Older cities exhibit higher densities toward their centers than do their younger counterparts. The modern urban development trend, then, has been toward increasingly lower densities over an entire area. It results, partly, from the development of the automobile-highway system of transportation and, partly, from rising standards of living and the desire for increased living space.

The density curves depicting economic activities are similar to the residential curves, though older cities may still have a higher density of economic activity at their cores. Still, suburban areas now offer more employment opportunities than do central city areas. These density curves indicate, again, the increased and accelerating consumption of land as urban areas grow. Since places of employment are generally the places people shop and seek services, the density trends for these activities are quite similar to employment density trends.

Examining both indoor and outdoor leisure time activities, we find the density curves for indoor leisure resemble the residential curves; the curves for outdoor activity show an opposite trend.

There is a significant difference between younger and older cities with respect to out-

Components: Population, Physical Structure, and Resources 89

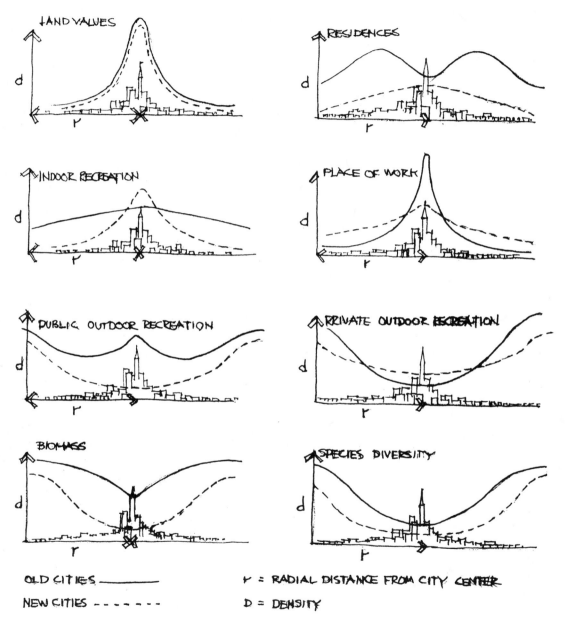

Figure 6 Major density patterns in urban ecosystems

door leisure activities. Younger cities tend to have less public open space than older cities. Los Angeles, for example, has the fewest acres of public park per capita of any major city.

Finally, density curves may also describe the plant and animal populations within urban areas. The relationship between younger and older cities in this regard seems particularly interesting.

In urban areas, the temporal cycles of activities and the resultant density cycles are often of more relevance to the functioning of the city than are "average densities". The nature of congestion problems appears to be changing. Traffic densities during the day, for example, are not as peaked for morning and evening rush hour, except perhaps those on mass transit, as they once were, although the overall level of traffic may be increasing. In fact, there are increasing levels of outward-bound morning traffic and inward-bound evening traffic in urban areas. The process of density smoothing, it seems, is now underway despite the fact that many congestion problems are still evident. But density smoothing could be improved, particularly through institutional changes.

Differentiation

Differentiation refers to the ways different sections of an urban area are used. Most simply, an urban area can be defined according to types of land uses and the spatial arrangement of those uses. Scale is as important a consideration in discussions of urban differentiation as it is with density.

Suburban areas are now more homogeneous than were their historic counterparts. Even in older sections of cities, residential areas are becoming more exclusively residential, and commercial areas more exclusively commercial. The neighborhood diversity of corner markets, small shops, restaurants and housing are being replaced by clusters of supermarkets, shopping centers, apartment complexes and entertainment areas. This pattern is even more evident in newly developed areas of the suburbs.

Reflecting the shift from parcel-by-parcel development to tract development, the areas of homogeneous land use are becoming larger and the start-up costs, planning scope, and infrastructure requirements are continually increasing.

Finally, differentiation of urban areas, as a whole, is increasing. Land is being used for a greater number of purposes, mirroring the increasing diversity and complexity of urban society.

On one hand, spatial differentiation in urban areas reflects more and more diversity; yet when examined at the small scale of the local area or neighborhood, land use appears to be more homogeneous. These homogeneous communities are expanding, producing what can be described as a less complex urban land use pattern.

Connectivity

Connectivity refers to the set of physical structures or functional linkages that tie an urban system together. In urban areas, structural connectivity is increasing at an accelerating rate. We have already pointed out that the accelerating consumption of land is causing urban sprawl. Urban residents, as a result, must travel farther to engage in various activities. And the activities themselves are becoming more spatially dispersed. Thus urban residents are traveling in more directions than in the past, when movement was chiefly towards the urban core.

Increased connectivity is at once a function of configuration, density and differentiation, and a stimulus to them. The building of a major highway, for example, stimulates a variety of other developments. The growth of urban areas thus can be described in terms of a positive feedback process that relates configuration, land use and density of activities with spatial access.

The maintenance of connectivity in a urban system requires a substantial input of energy and materials. Part of the "energy crisis" stems from the nature and patterns of connectivity in urban areas; the major impact of urban transportation on energy consumption, however, is indirect. About 24 per cent of the energy consumed in the United States is used for transportation; at most, only 12 per cent of all energy is used for all urban transportation. But 74 per cent of the population lives in urban areas; in an overall accounting, therefore, they consume at least 74 per cent of our total energy. The relatively small proportion of energy consumed for purposes of urban connectivity, thus, is only one of many factors supporting the urban life-style with its high energy consumption.

There are other forms of connectivity than those relating to the mobility of urban residents. The physical structures of utility systems—electricity, gas, water, and telecommunications, for example—all require some energy input, though at considerably different levels. Where these physical systems perform similar functions, it may be possible to decrease energy use. A shift in emphasis from mobility to telecommunications, for example, may seem to result in more efficient use of energy. But this shift to communications may actually increase energy consumption by increasing the geographical areas of cities, the sizes of enclosed living spaces, the need for wastewater collection and treatment systems, etc. Urban systems are complex!

Because urban development has a strong positive feedback through connectivity, intervention in the planning of urban linkages can be used to guide the development of urban areas. Institutional structures for connectivity, which operate at a larger scale than do institutions responsible for other urban structures, can thus play a significant role. Intervention at this level can have far-reaching results and has the potential for harm, perhaps, as well as for good.

We have thus far provided some notions about why cities exist and we have used the concepts of configuration, density, differentiation and connectivity to describe the characteristics of cities. These suggest certain issues that should be kept in mind in the examination of urban problems and the search for solutions.

First, the city as we know it today reflects the processes, technology, resources and human aspirations of the past. Likewise, as a physical artifact, it is a record of the past. The processes and attributes affecting urban growth have changed over time; they are changing now and will continue to change. At issue is the compatibility of current and future societies and the physical urban situation they inherit. New modes of transportation, for instance, are of concern. Not long ago, the average speed of travel in the city was dictated by the speed of a man's walk or of horse-drawn vehicles. Average speeds ranged from three to five miles per hour. Development of the horse-drawn railcar, the streetcar and the subway raised average speeds along routes radiating out from the city. Then came the automobile, not restricted to radial routes, which raised average speeds to as much as 30 miles per hour. But the lifetime of the physical structures in cities is measured in many decades while transportation technologies have been changing every few decades. Transportation technologies are constrained by the necessity to function in the older physical urban environment.

Secondly, the resources available to, and the aspirations of, urban dwellers have changed and new arrangements have been sought for the layout of residences, offices and recreational facilities. The physical urban structures built in previous eras are outmoded in terms of present resources and perceived needs. Further, different segments of the population have different resources available to them, and different aspirations. Between groups, there is constant competition for space taking place in an inherited urban environment which cannot meet all current needs. Clearly, one source of urban problems is the mismatch of the physical urban setting and the present aspirations and activities of those who live and work in that setting. Needs, too, are constantly changing—we must move faster and faster to keep up. Not only must we resolve the conflicts between what we have and what we want, but we must look forward to what our desires will be in the future if we want to move the urban environment closer to what we will be able to use.

Third, while the city exists in one place or site, it is the product of a number of processes, many not confined to the city's geographical area. In natural ecosystems, these processes are constrained by the site and its resources. The density of organisms, for example, depends on the food and resources available to the population in the area. In nature, the development of greater connectivity at a site results in increased efficiency in energy and materials use. In the city, however, resources are not limited to those at the site. Patterns of diversity and connectivity reflect, partly, the way a city relates to its larger situation or environment rather than to its local site. As a result, when we consider land use patterns or other urban problems, we must look beyond the geographical confines of the urban area to the regional setting and examine how the city interacts with outlying areas and how it competes with other cities.

FUNCTION OF THE URBAN CENTER

The conventional method of describing the way a city functions is to describe the ways urban residents spend their time—at work, at play, at home, etc. We will use this conventional approach in our discussion, but we recognize there are other ways to think about the functions of the city. One approach is to view the city in relation to its hinterlands; its functions then can be seen as information and materials collection and dissemination, health and government services, etc. We might also classify the functions of cities as trade, governmental, recreational, etc. Viewing the city as a home for man, we might examine the way a city provides for fundamental human needs—food, shelter and security, for example. In this discussion, however, we will describe urban functions conventionally, according to the major activities of urban dwellers: work, habitation, recreation and transportation. These activities are not mutually exclusive and our approach to them incorporates aspects of other approaches.

Work

Work is one of the urban dweller's most important activities. Workers can be classified in several ways. For example, we may distinguish workers in "basic" activities from those in "service" industries. The basic industries, those producing material goods, provide products to be exported from the city, enabling the city in turn to import the goods it needs to survive. Service activities are those which support basic activities. A worker producing

Egyptian woodwinds to be exported and sold throughout the world is employed in a "basic" industry, while the person who washes the woodwind maker's clothes is employed in a "service" industry. Though the ratio of basic to service activities varies according to the size and type of city, each job in a basic industry usually generates one job in a service industry. At the present time, however, service industries appear to be growing more rapidly in our economy while employment in the basic industries, such as manufacturing and agriculture, has been declining for several decades.

Classifying industries as "basic" and "service", however, presents a problem. These activities greatly depend upon each other. A worker in an iron or steel mill produces items primarily for export, but his industry's competitive advantage depends partly on how well the service industries perform. Further, certain industries might be considered both service and basic; the functions they perform may be services, which also constitute the city's principal export.

Service industries can also have a major influence on regional or nation-wide growth. Many developing countries, concerned with the concentration of their population in only a few urban centers, have tried to induce urban growth in regional centers by attracting or locating basic industries there. But this policy has proved disappointing. In Venezuela, for example, Caracas has continued to grow rapidly despite efforts to promote growth elsewhere. The service industry in Caracas is so well developed that outlying regions, despite inducements to industry, are unable to attract major segments of the population or to serve as centers for urban growth. Promoting basic industry is useful in developing an area, but more needs to be known about the interdependence of service and basic industries and of the mechanisms involved.

The rewards for working are many and varied. Societal interests, as a whole, are served by an individual's commitment to work. The individual may derive some satisfaction from this knowledge. But on a more tangible level, the worker receives a monetary reward for his efforts, enabling him to buy goods and services from others.

There are sometimes psychological or motivational rewards for working. Though some activities are boring and monotonous, others reward the individual by giving him a sense of participation and accomplishment. A major challenge to urban society is to structure work activities so that every individual is broadly rewarded for his participation.

Each type of work activity has characteristic requirements for buildings to house the operation, streets and railroads over which to transport materials and products, communications facilities and disposal systems to carry away waste products.

Work activities using physical processes—construction, for example—require a high concentration of energy and materials and rely heavily upon technology. Such concentration of energy and materials also produces environmental pollutants. As a result, sophisticated technology, government, and other mechanisms have been developed in the city. Continuing technological and marketing changes in our society have meant that the work activities requiring heavy concentrations of energy and materials are continuously changing as well. Anticipating, building for and accommodating such changes are significant challenges for any city.

Habitation

The term "habitat" describes the area or space provided to meet a population's requirements for living. One limited interpretation of urban habitat is residential or living area, which is typically described in terms of structural type and related density capacity—from low density single family units to medium density multiple family units to high density high-rise apartment units. Though such a classification seems acceptable, the relationship between density and residential unit type is not a simple one. Density refers both to the number of persons per unit of enclosed residential space and to the number of persons per unit of land occupied by residential structures. Suburban single family units may have a low density in terms of residents per unit of area, but not in terms of residents per unit of living space. In fact, suburban areas occupied by young families typically have very high densities per unit of living space. On the other hand, high-rise residential units, when occupied by wealthy families, may have high densities per unit of land, but relatively low densities per unit of living space.

Habitation for city residents includes not only living space, but also work, recreational and educational space, etc. Any broad evaluation of an urban habitat must examine this complex of environments, giving attention to the freedom of movement, protection and security they provide. It must examine also how well these environments support the individual's diverse needs for stimulation and interaction at some times, for privacy at others.

Human and animal habitats sometimes appear analogous; territoriality and human habitation needs are often compared. Such parallels are strained. Cultural influences on man's behavior and the extreme complexity of his living situation make it difficult to determine accurately man's minimum territory or living space requirements. Diversity of habitat may be essential for man, making it difficult to assess both the kinds of spaces required and the minimal requirements of those spaces.

It is, nonetheless, widely acknowledged that parts of many major urban centers, both here and abroad, are overcrowded. Whether it is desirable to provide greater diversity or more living space to ameliorate overcrowded conditions is a question which remains to be answered.

Providing for human habitation necessitates adequate life support systems. Residents must have sufficient food and water, clean air, waste disposal and protection. Such systems require physical facilities: 1) food—physical production and transport facilities; 2) water—water supply, treatment and transport facilities; 3) air—air pollution control facilities; 4) waste disposal—waste water and solid waste disposal and treatment facilities; and 5) protection—controlled temperature facilities, shelters and structures.

Man has relied heavily on technology to support human habitation. In wealthy nations, technological expertise, materials and energy have been used to override the physical characteristics of an urban site. Buildings have been constructed with apparent disregard for exposure, soil conditions, wind loadings, drainage, earthquake conditions, or other characteristics. These matters are not ignored, however; they are simply circumvented, to a large degree, in the design process. The designer, for example, might increase the heating and cooling capacity of a building rather than rely on the building's orientation to the sun.

Habitation includes open spaces as well as sheltered structures. The recent trend in the development of American cities has been to include more private open space in the basic habitation unit, the residence. Even high-rise and multiple family units tend to have more open space around them than in previous decades. Likewise, suburban homes are being built on larger and larger lots. This trend probably reflects the greater affluence of the population, which allows them to substitute private for public open space.

Recreation

Recreation relates man's pursuit of pleasure to relaxation; for many, recreation is a diversion from work-related activities. Recreational activities may be classified as either indoor or outdoor recreation. Outdoor recreational activities can be generally categorized according to the areas where activities are located:

Area of Recreation	Time Spent at Activity
Neighborhood	15 minutes to a few hours (daily)
City	Few hours to one day (periodically)
Suburban	Few hours to weekends (periodically)
Regional, State or National	Weekends to weeks tourist recreation (annually)

Facilities for recreational use range from those specifically built for recreational purposes to those intended or designed for other purposes but informally used for recreation. Man-made facilities such as playgrounds and amusement parks and natural facilities such as lakes and wildlife refuges are all recognized and identified as recreational areas in land use surveys and analyses of urban communities. Streets, alleys and vacant lots serve as recreational facilities but are not identified as such in land use surveys. They do support a significant portion of urban recreational activity. The nature and quality of these facilities are generally related to the physical characteristics of the high, medium or low density developments where they are located. High density developments, for example, may stimulate games like street hockey; sand lot baseball is more often found in medium or low density areas.

WE ALL USE THIS MARSH FOR...
- FLOOD ABSORBTION
- WATER STORAGE
- WILDLIFE HABITAT
- SCIENTIFIC STUDY
- ENJOYMENT AND MORE..

Increasingly, urban residents are leaving the city in search of outdoor recreation. The urban system encourages the use of public and private campgrounds or cottages and farms owned by urban dwellers, though these facilities need not be located in the urban system itself. Those with the necessary resources can travel long distances to engage in recreational activities. The burgeoning market for camper-type vehicles, for example, reflects the urban dweller's expanding horizons and utilizes a new form of mobile recreational space.

Many cities intend to maintain their present green areas and to create new ones to make outdoor recreation and semi-natural environments available to all their residents. Such urban open spaces, however, do more than simply provide for recreational needs; they also provide natural drainage and peak flow storage, rather than expensive man-made drainage systems, and can help ventilate and moderate temperatures in urban areas. Less has been done in the United States than in other countries where green spaces have been included more successfully in the urban environment. Lack of appropriate action in this country may be the result of the alternatives available to urban dwellers here—they can have their own private outdoor space by buying a large residential lot, or they can travel if they wish. It may also be a result of marketing and trading mechanisms unsuitable for the creation of commonly owned green space.

Presently we have no practical mechanism for translating the urban dweller's need for public outdoor green space into economic terms to compete with the need for housing or commercial development on expensive urban land. Such mechanisms do exist in theoretical form. Some work has been done on computing the value of recreation based on the amounts people are willing to spend for travel or for recreational activities and facilities. A mechanism to convert such willingness to spend into the hard cash for recreational land is presently lacking.

Transportation

Transport activities are primarily support functions moving people, materials and information within cities and between them. Only rarely is travel an end in itself, e.g., the Sunday drive. Transportation can be categorized according to what is being transported—people, goods, information, utilities or services.

Many physical means of transport—elevators, autos, trucks, mass transit, railroads, transmission lines and pipelines—date from the turn of the century or earlier. Some are vehicle and guideway combinations such as railroads and tracks or motor vehicles and highways. The movement of natural gas, fresh water and waste water through pipes represents another basic form of transport, and the continuous belt escalator still another. Change in each of these forms has been relatively slow, new modifications coming over decades rather than months or years.

Communication has been included in the transport category because it is similar to the others in function and form. There are, however, two notable differences between communication and the other transport activities. First, except for letters and books and other physical types of messages, the physical structure requirements for communication are few. Communication devices can, and have been, substituted for physical movement in the urban society.

Second, communications forms, like other forms of transport, have recently been subjected to radical changes. Presently, changes in communications are stemming from developments in television, such as the expansion of cable television systems which can handle

many channels of information flow and have yet undeveloped potential for two-way communications.

Another significant development in information transport, the linking of computers to communications systems, has had considerable impact upon the patterns of production activities such as fiscal or inventory management. The effect of this development on the urban way of life has yet to be determined.

Some may dismiss these developments in communication as unimportant. Others, however, argue that the city is undergoing a communications revolution and feel that the challenge to urban society is to understand the change and to mold the new technologies into constructive forces. Because transport and communications influence almost every facet of urban life, they should be regarded as "high-leverage" technologies. Any change in their performance characteristics or in policies affecting their organization and control could have far reaching effects.

One-fourth or more of urban land is devoted to transportation, particularly of people and goods. These areas of transport are frequently unsafe, congested, noisy and polluted. Vehicles spread dirt and grit and burn fuels which create emissions problems. Conflicts between motorists and pedestrians, between highways and adjacent lands, and between vehicles themselves have received much attention in the past; somewhat less attention, however, has been given such issues as the effect of transportation forms on the shape of the city and the urban environment. Transportation is a key facet of the urban environment; to examine only its superficial problems will not reveal its social importance nor permit formulation of appropriate policies for future development.

Transport, serving to organize the urban center, is somewhat parallel to the energy,

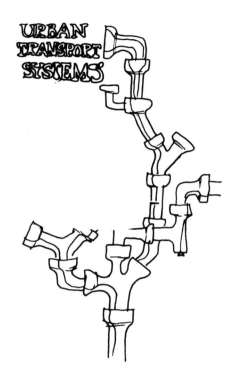

materials and information flows giving organization to natural ecosystems. In natural systems, however, transport is accomplished through material cycles which become tighter as the systems develop. More internal cycling evolves; there is less exchange outside the system and there is little surplus. In urban transport, on the other hand, there is an ever-increasing and wasteful energy flux through the system as it evolves.

Energy use in transport, though, is a complex situation and must be viewed in terms of the total output of the urban system. The question then is whether or not the urban system is producing as much as it can, given the amount of energy it is consuming. Some critics look at the energy issue in terms of small cars versus big cars, one mode of transportation versus another, or reorganization of the city so less transport would be necessary. These seem reasonable approaches, but they often fail to consider the many tradeoffs involved. The city provides its residents with an attractive variety of choices. Putting work and recreation areas close to home, however, and possibly reducing the city to the village level, may restrict an urban dweller's choices. Loss of choice and loss of productivity resulting from decreased energy use, therefore, become paramount concerns. We might ask also how readily accessible energy might be used to provide more amenities locally.

RESOURCES: WATER, ENERGY AND MATERIALS FLOWS

Urban populations consume goods, services and energy, and in the process deplete natural resources and generate wastes. In the past, man has had only slight impact on the environment. Today, however, the world's population and technological prowess combine to consume resources and spill wastes into the environment at an alarming rate.

Our energy demand is increasing exponentially, and at the same time, we are rapidly exhausting other resources as well. As resource consumption increases, so does the rate at which we discharge wastes into the environment. The detrimental effects of these residuals are well-documented. The problems of resource consumption and residual discharge are nowhere more concentrated and acute than in our cities.

The system in which energy and materials move through the urban environment, driven by the demands of the human population for goods, services and energy and powered by fossil fuels, consists of three main segments: the man-made energy and materials pathways; the natural environment; and the socio-economic-political system controlling the process.

The Man-Made Materials and Energy Flow System

The system in which energy and materials move extends from the extraction of resources for manufacturing and consumption to the discharge of wastes into the natural environment (Fig. 7). The production sector of the materials and energy flow system encompasses all the facilities that produce goods, services, and energy to meet the demands of the human population. It includes both corporate and government activities in extraction, generation, processing, and transportation. All material entering the production sector is ultimately discharged, either as a useful product or as waste.

Products supplied to the population include goods, fuel, food, services, electricity and water. Some of the goods manufactured may be designated personal capital—they are used, rather than consumed, until they are worn out or obsolete. All other products are consumed and then discharged in some fashion. Residuals from both production processes and consumption of products can either be discharged to the environment—with or without treatment—or recycled into the production process.

The Natural Environment

Material and energy flows in the man-made system affect natural systems in three ways: stocks of nonrenewable resources are depleted and renewable resources are over-utilized; re-

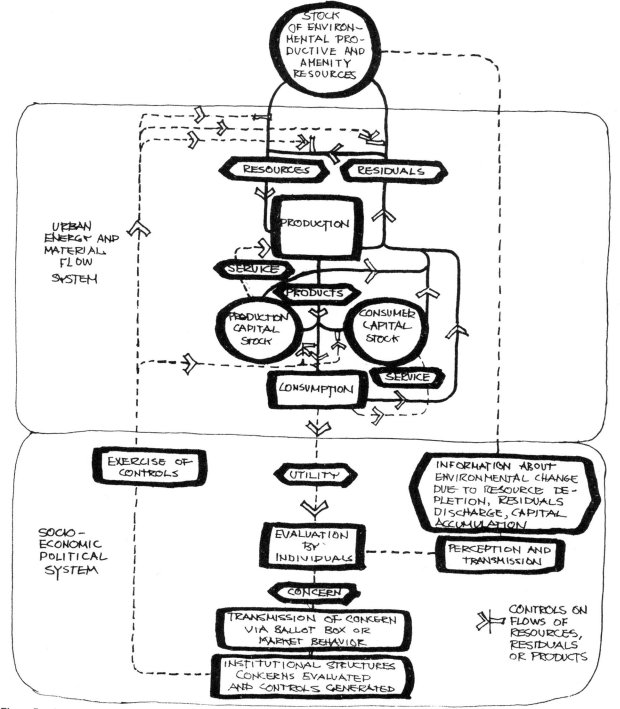

Figure 7. A conceptual system of energy and materials flow

sidual discharges can produce adverse changes; and part of the natural environment is preempted by man-made structures and man-controlled spaces for the production and use of goods, services and energy.

Resource Depletion and Over-Utilization

Resources, as discussed here, are materials that are limited because of their amount or distribution in time or space. The Committee on Resources and Man (1969) defined resources as that proportion of the total stock that can be made available for human use under any present or possible technological and economic conditions. "Total stock", on the other hand, is the sum of all components in the environment that would be resources if they could be extracted. Resources include renewable and nonrenewable energy sources, water and minerals.

A man's basic energy requirement, for food, is about 2,000 kilo-calories per day. Each person in a highly developed industrial society requires more than 200,000 kilo-calories per day for support. This energy demand continues to increase exponentially.

The current rate of energy consumption in highly developed countries, the fact that most of the world's population still consumes energy at minimal levels (200,000,000,000 people still have no electricity), and the fact that population in both developed and underdeveloped countries is still increasing, taken together, indicate the immensity of potential energy demands.

Available fuel supplies are limited and becoming increasingly inaccessible. Of the readily available fossil fuel a significant portion is located in politically sensitive and economically underdeveloped parts of the world. New sources of energy will probably not be available much before the beginning of the next century. Increased energy demands in the face of limited supplies may result in significant price increases. Increasing fuel imports—it is estimated that the portion of the U.S. energy demands supplied by foreign sources will increase from the current 12 per cent to 30 per cent in 1985 (Malin, 1973)—raise questions of national security, morality, balance of payments and the effects on domestic fuel interests. Increasing development of domestic fuel sources, on the other hand, raises environmental and health questions associated with the possibilities of increased strip mining, relaxation of air pollution standards to accommodate more polluting fuels, and problems of radioactive waste disposal and thermal pollution, inherent in the development of nuclear energy. The need for maximum efficiency in the use of energy resources is readily apparent.

Of all resources, water is the most necessary and the most available. In the natural environment, water is needed to sustain life; it is likewise necessary for man's survival. It transports elements, carries away waste products and provides man with recreational opportunities and aesthetic satisfaction. As a material resource, water is used heavily in the production of goods and services. It is a renewable resource as well, since it is the vehicle for obtaining energy from the movement of tides and rivers.

Because of its unique characteristics—its ubiquity and its role as both a material and energy resource—water deserves special consideration. Water is not uniformly distributed in time and space. Its availability as a material resource is further limited by the effects various uses of water have on water quality. Continuing to supply clean water to large populations in water-deficient areas and the further development of hydroelectric energy at significant expense to the environment are likely to create increasing economic and political problems. These and other prospects indicate clearly the necessity for the conservation of both water and electrical energy.

"The amount of metal consumed in about thirty years, at the current rate of increase in consumption, approximates the total amount of metal used in all previous time" (Committee on Resources and Man, 1969). Cloud (1971) has estimated that of known recoverable U.S. reserves for 17 mineral commodities, 11 will be exhausted by 1985, 15 by the year 2042. By then, the world reserves of ten of the 17 will also be exhausted. Substitutes for some minerals can be found if there is sufficient economic incentive. "Some mineral resources, however, such as helium, mercury and uranium or thorium have unique properties and for those there are no satisfactory substitutes" (Committee on Resources and Man, 1969). With scarcity will come increased prices. Should an inexhaustible source of cheap energy become available, it still could not surmount the labor and capital costs required to extract widely dispersed minerals from massive quantities of rock. It is clear, therefore, that we must find ways to reduce the input of critical materials into the production process.

Residuals Discharge

Residuals are the "leftovers" from human production and consumption activities. Whatever energy or mass is not incorporated into a product or is left after the consumer has used a product is a residual. These substances may be waterborne, airborne, or solid. Residual energy is commonly discharged as heated air or dissipated into the atmosphere by radiation or convection. Residuals have zero prices—prices below their variable costs of production in present markets (Russell and Spofford, 1972).

The environmental effects of residuals discharge, including threats to human health, have been well-documented and require no elaboration here. Residuals of particular concern include: liquid wastes from industrial processes; refractory substances, including biocides and heavy metals; air pollutants, particularly from transportation; nutrient discharge in domestic wastes; and heated water and radioactive wastes from power generation. Residuals scoured from urban and agricultural lands by runoff from rainfall represent particular problems in water pollution control. Runoff may carry biocides, nutrients, minerals and debris from the land into our lakes, rivers, streams and oceans.

Utilization of Space

Man's total physical environment consists of the natural environment together with those parts of it man dominates and the structures he has accumulated. Man-dominated space and structures are devoted to the extraction of raw materials, the generation of food and energy, the production of goods and services, the disposal of wastes and the housing and care of people. Increasingly man's various activities compete for the remaining underdeveloped areas of our finite globe. Natural systems are the passive and, often, losing competitor. Thousands of square miles are inundated to tap river systems for hydroelectric power. Thousands more will be demanded for the extraction of uranium, coal and scarce minerals. Waste disposal and storage will require further space. The structures required to house and transport man, and to manufacture the products he demands, cover the earth.

"The most obvious and acute pressure upon space is in our great cities and surrounding metropolitan areas whose existence and expansion depend greatly upon technology. They, and the associated industries and highways that connect them, are absorbing agricultural land in the United States at the rate of some million acres a year. This means fewer orange and walnut groves in California, dairy farms in George, truck and tobacco lands in Connecticut and less of the proverbial fertile valley along the Miami in Ohio" (Sears, 1958). "And between cities, across the land, highway departments are busily freezing the nation into a permanent interurban geometry. Often, in fact if not in theory, they are responsible to no one but themselves and their Euclidean rule that the shortest distance between two points is a straight line" (Sears, 1958). "If the electrical industry's projected quadrupling took place in the next 20 years, their transmission lines would require an additional 3,000,000 acres or 4,700 square miles, equal to the entire area of Connecticut" (Stein, 1971).

Man-dominated space and man-made structures could well exceed the ability of the earth to physically sustain them. They seriously encroach upon the untouched natural systems man relies on for recreation and aesthetic satisfaction. Unfortunately, the poor quality of much of man's work in the space he dominates and the structures he builds means that many people live in ugly, depressing or unhealthy surroundings.

"To consider whether the earth might support three doublings of the human population is probably a purely hypothetical situation. It seems more likely that further crowding, the necessary social and governmental restrictions that accompany dense settlement, and certain kinds of boredom resulting from isolation from nature in an immense, uniform, secular society may prove so depressing to the human spirit or so destructive to coherent social organization that no such population size will ever be reached" (Committee on Resources and Man, 1969).

Socio-Economic System

The socio-economic system is the primary control on the movement of energy and materials through the man-made system. Changes in the environment from resource extraction, residuals discharge, or the number or nature of physical structures, are perceived by individual consumers. Consumers weigh decreased satisfaction resulting from environmental change against increased satisfaction to be derived from goods and services.

The process of evaluation is represented graphically in **Figures 8, 9,** and **10**. To illustrate

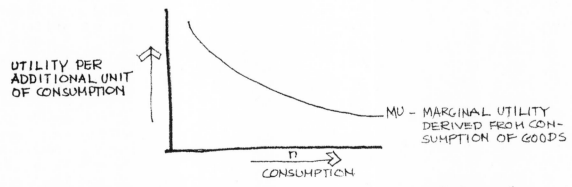

Figure 8. Consumers obtain decreasing amounts of utility (satisfaction) from each succeeding unit of consumption.

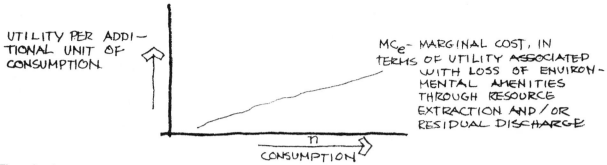

Figure 9. As consumption increases and resource extraction and residuals discharge rates increase concomitantly, each succeeding unit of consumption entails greater environmental cost.

a point, homogeneity of materials output and of environmental resources (both material and productive and amenity resources) has been assumed. From the principle of diminishing marginal utility it can be argued that the greater number of units of consumption (assuming a given level of population and a fixed distribution of material wealth), the less each additional unit of consumption adds to the total utility or satisfaction of the population per unit of product consumed (Fig. 8).

Increased consumption, requiring increased production, presumably requires increased rates of resource extraction and residuals discharge with attendant degradation of environmental productive and amenity resources. With increased consumption, the satisfaction from or the utility of the environmental resources declines. If consumption is increased, an environmental cost is incurred that can be expressed in terms of lost utility (Fig. 9).

Evaluation of the environmental costs of increased consumption depends upon the accuracy with which both short- and long-term environmental changes are perceived. It also depends on how efficiently those perceptions

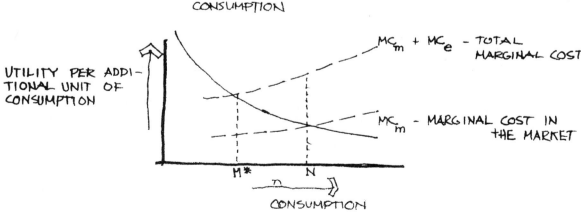

Figure 10. Market forces will dictate consumption to point N where marginal cost equals marginal utility. But if total costs are considered, consumption will occur at level M*, where total satisfaction is maximized.

are transmitted, first to the individual, and by him to the political-socio-economic system where it is evaluated in light of other priorities and pressures. Then action is taken, if necessary, to bring about change in the materials and energy movement system. Whether, and how, materials and energy flows are altered in response to resource depletion and residual discharge depends on how effectively the whole system functions. The process is subject to the lags, thresholds and historical properties that have characterized urban systems (Holling and Orians, 1971). Furthermore, the individual finds himself weighing products against environmental resources outside the market system without knowing fully the consequences of his choice in terms of environmental effects. He has no assurance that any action taken on the basis of his evaluation will bring about any desired improvement.

Figure 10 illustrates the effect of failure to completely evaluate environmental resources when decisions are being made that affect consumption levels. If the value of both products and environmental resources lay wholly within the market system, material goods would be consumed at the point where marginal utility equals marginal cost and total utility is maximized. However, since the value of environmental resources lies, for the most part, outside the market system, there is an inherent tendency for consumption to occur at the expense of environmental quality and total social well-being.

A CONCEPTUAL VIEW OF THE URBAN SYSTEM

We cannot place a boundary around the urban system to enclose it. The materials and energy flow system supporting the urban population extends from the source of productive

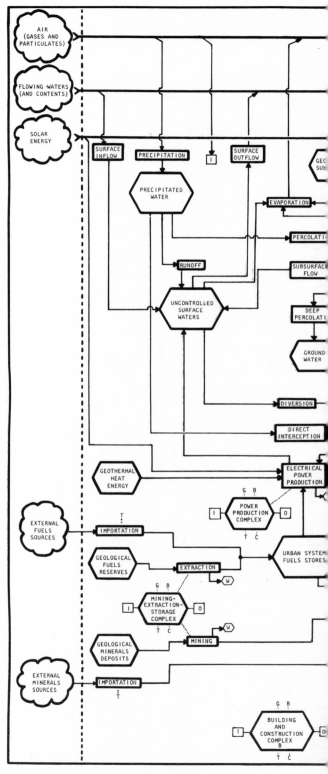

Figure 11 Urban energy and materials flow system

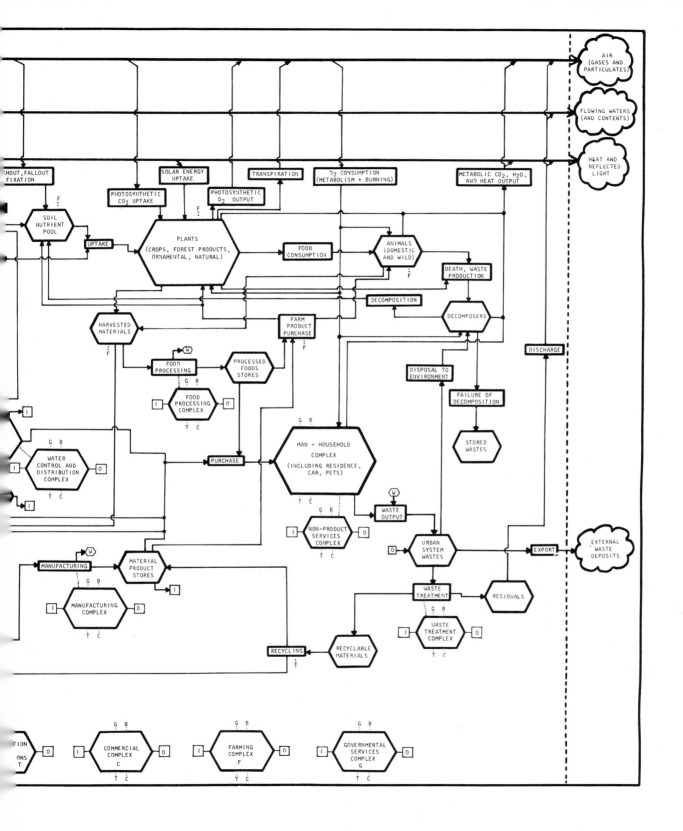

resources to the discharge of residuals into the environment. The distinguishing features of an urban system are people, their personal capital, and varying amounts of productive capital. Other aspects of the urban system are not necessarily within the system itself. Productive resources may be extracted miles or even continents away. Production of many goods and services may take place beyond the city limits. Much of the capital and the sociopolitical system supporting the community is frequently located elsewhere.

A more comprehensive view of an urban energy and materials flow system than was shown in **Figure 7** is indicated in **Figure 11**. Its flows include both the natural and the man-made systems. The urban system centers on the set of human households that exists in the city itself; it extends, on the one hand, to the initial points where energy and materials flow into the system and, on the other, to the effective impact limits of the waste or residual outputs. The diagram does not describe the spatially-defined urban area, but the entire system of urban support and impact.

We have attempted to integrate into the diagram the fluxes of matter and energy occurring through biological processes and human economic activity. Three major through-flows of the natural environment are assumed to exist: 1) the flow of air with its various gases and suspended materials; 2) the flow of water with its dissolved and suspended materials; and 3) the flow of radiant energy. The system takes materials and energy from these flows and returns material and energy to them.

The economic sectors of the diagram allow basic inputs of raw materials and fuels from outside the system and the exportation of wastes to locations outside the system as well. The remaining resources—including ground water, geological fuels and energy, mineral resources and all foods and organic raw materials—are assumed to be obtained internally.

Connections between the non-living environment, biological processes, and the human sector are represented by various water, nutrient, mineral resource and energy pathways. The primary connections with the physical environment involve: the use of water and nutrients in biological production processes; the use of water, minerals, fuels and organic food energy by man; and the output of the organic and inorganic wastes of human activity into the environment.

The organization of the diagram is intended to account for the entire complex of activities involved in the functioning of human households. Accumulations of matter, energy and potential human services are represented by hexagons. The processes resulting in the delivery of energy, materials or services, like food processing or power production, are designated by rectangles. Aggregated hexagonic figures represent economic complexes having significant capital accumulations of materials and resources and flow-throughs of material and energy for maintenance. Theoretically, each process carried out by man has a complex associated with it. In our scheme, many specific complexes are aggregated into larger ones, to which they are related in terms of space or function.

The substances flowing through the systems are categorized as inputs (I) of water, electricity, fuels and products; outputs (O) of wastes stemming from system maintenance; and wastes (W) resulting from manufacturing or the production of goods. The two aggregated waste categories consist of gaseous, liquid and solid wastes, as well as waste heat from power and fuel consumption. Both kinds of waste enter a general urban system wastes pool.

Several economic complexes are included in

the diagram. Others, because of their complex involvement with many processes, have been shown at the foot of the diagram and their points of involvement have been indicated by abbreviations. Processes acting to control flows and costing money are indicated by the term "purchase", whether they were actual purchases or were paid for by taxes.

Our diagram, of course, is incomplete. It does, however, help identify many of the points at which energy and materials flow processes are, or can be, manipulated. The consequences of such manipulations can be qualitatively identified to a limited degree. For example, it is apparent from the diagram that reducing the export of wastes to locations outside the urban ecosystem, while holding waste treatment operations within the system constant, will inevitably lead to increases in direct waste disposal and an increased accumulation of residuals within the urban system. Likewise, it is apparent that by mass balance decreases in the rates of materials flowing through the system will affect a large set of relationships, both at the resources extraction end and at the wastes output end.

SUMMARY

The primary components of the urban system are: 1) its human population; 2) its non-human populations; 3) its physical structure; 4) its function; and 5) its resource flows. They are intricately interrelated and depend greatly upon each other. The urban population, and providing for it, are the city's reasons for being. Without individuals to work in various capacities, the urban system could not function. Without transportation to move the individual from home to job to recreation, without structures to house people, production activities and recreational facilities, survival in the city would be impossible. Without the necessary material and energy resources, urban life would come to an end. Ideally, all the urban components work together for the survival and enhancement of human life.

The components and the interplay between them, however, present various problems. The number, density and heterogeneity of the urban population, for example, affect all aspects of city living. Overcrowding not only may create stress situations for urban residents, but also may strain the capacity of the urban system to support its population. The psychological effects of high human population density, though not entirely clear, seem to lead residents to compartmentalize their interests, behavior and attitudes. Much work still needs to be done to explore the ramifications of high population density.

The factors determining human population size are fertility and mortality rates and migration. Birth rates are the most difficult to project. Changes in fertility rates depend on a number of factors. Changing social attitudes affect family size; biological and economic factors are considered important as well. Accurate determination of human fertility trends is necessary if we are to adequately assess future population size and the resulting economic and resource demands.

The human mortality rate plays a relatively insignificant role in determining urban population size.

Migration is often the most important influence on urban population size. Individuals between the ages of 20–32 and 0–4 (their children) are those most prone to migrate; longer residence in a region decreases the likelihood of migration. Both ethnic grouping and education are important factors in migration patterns.

In the past, systems constructed for modeling migration have been limited in scope. A broad, interdisciplinary model of movement, encompassing economics, politics, climate, geography and population density, etc., would be useful for projecting interregional migration. Because of the enormous affect immigrants have on urban areas, accurate knowledge of migration is essential to understanding urban population processes.

The distribution of human populations in our cities according to age, ethnic, and socio-economic characteristics creates problems in providing adequate housing, education and job opportunities for minority races and classes.

Clearly the problems of population density and distribution are not easily solved. Density and distribution, fertility and migration create population pressures that affect all aspects of the urban ecosystem. If we are to survive in the city, we need to understand these pressures and act wisely to relieve them.

Urban populations include more than the human element. Other organisms—mammals, birds, reptiles, and insects, etc.—are also found in the city. These species both provide enjoyment to man and create problems in the city. The study of urban fauna and flora, however, has been badly neglected.

The two major threats from urban animals are disease and animal-bite. Many animal pathogens can cause disease in man. Animal-bite is a major source of injury in the city. In addition, animal and bird fecal materials are often sources of disease.

Urban plants are not as damaging as urban animals. Plant allergies and accompanying respiratory diseases do, however, constitute sources of discomfort, illness, and sometimes death in the city.

Urban plants and animals bring psychological and economic benefits to urban residents. Nature in the city brings man pleasure. The cost of feeding and maintaining urban species results in large and direct monetary exchanges included in the Gross National Product. Both plants and animals can serve as indicators of the health of the urban environment. Both contribute significantly to the aesthetic and cultural enrichment of human life and serve to improve the habitability of the urban environment. They are important facets of the urban ecosystem and need to be better understood if we are to form ecologically sound policies.

The physical characteristics of the city are largely the result of the time of its most rapid growth, its site, and its strategic location. Its history, physical properties and relationship to other cities and resources are significant influences on its present condition. These characteristics underlie the city's configuration, or general shape and pattern; its density, or population per unit of space; its differentiation, or the way various parts of the city are used; and its internal connections, the pathways linking the parts of the urban area. Taken together, such characteristics and concepts describe the city's nature and structure.

The function of the city is defined in terms of work, habitation, recreation, and transportation and of the interplay between them. These four interrelated functions are essential elements of the urban system.

The system in which energy and materials move through the urban environment is driven by the demands of the human population for goods, services and energy and is powered at present by fossil fuels. It consists of three main segments: the man-made energy and materials pathways, the natural environment, and the socio-economic-political system which controls the process.

Our energy demand is increasing exponentially. We are depleting and over-utilizing material resources as well. At the same time, we are discharging waste products into the environment at an alarming rate. Man-made structures and man-dominated space may soon exceed the ability of the earth to sustain them; they encroach upon the natural areas man relies on for recreation and comfort. The problems of resource consumption and residuals discharge are nowhere more concentrated than in our cities. Water, energy and material flows require our urgent attention; without adequate flows of these resources, the urban way of life cannot be sustained.

Improving the relationships between the city's structure and functions and intervening in its materials and energy systems require that we accurately assess the condition of the urban system and act appropriately. Indicators of urban ecosystem health are essential. Models to abstract aspects of the system for purposes of analysis, problem-solving or design are necessary. These indicators and models will prove useful in the formulation of policies essential to the maintenance of urban ecosystem health.

RECOMMENDATIONS

Policy Recommendations

II-1. We recommend that enclosed cities, megastructures, ocean cities and other innovative settlements be considered primarily as experiments, not as panaceas for our urban dilemma.

II-2. We recommend that the Gateway proposals of the National Park Service receive more support and direction toward providing needed recreational space to inner city residents.

II-3. We recommend that funds to control urban pests be allocated to reduce the actual burdens of plant-induced and animal-transmitted diseases in the urban system; for example, when a disease is rare, funds spent to improve treatment and diagnosis may be more effective than those spent to control the source.

II-4. We recommend that researchers, policy-makers and funding institutions be required to look beyond the urban area to its regional setting and take into account the interactions of the city with outlying areas and other cities.

II-5. We recommend that structural and functional operations of urban areas be planned and organized around ecosystem units.

II-6. We recommend that the environmental resource needs of man-made urban systems be fully recognized in planning and that development proposals incorporate provision for these resources on a sustaining basis.

II-7. We recommend that national and state urban growth policy recognize and consider ecological sensitivity and capacity in matters of population allocation and urban design.

II-8. We recommend that environmental determinants be applied to the physical use of land. Natural components can be compatible with the physical uses of urban lands if the inherent physical and biological constraints on and opportunities for specific land uses are taken into account.

II-9. We recommend institutional changes which will reduce hourly, daily, and cyclic variation in the number of people concentrated in specific parts of the city.

II-10. We recommend that urban transport and communications be regarded as "high leverage" technologies deserving especial care in application; changes in their performance characteristics or in policies affecting their organization and control have far-reaching effects.

Recommendations for Action

II-11. We recommend the preparation of resource publications on urban wildlife and vegetation.

II-12. We recommend:
a) Support of participatory gardening projects in the city as a tool to encourage neighborhood/community cohesiveness.
b) Alteration of dog control programs, including leash laws, to eliminate free-ranging dogs.
c) More extensive use of vegetation in controlling air, noise and visual pollution in the city and in conserving energy used in heating and cooling urban structures.
d) Control of rats through urban renovation programs; efforts should insure adequate sanitation and physical conditions of human dwellings, since these are the most effective means of controlling rat populations.
e) Developing building design criteria to provide structural controls for bird populations.
f) The inclusion in public health records of more complete information on animal-transmitted and plant-induced diseases.
g) Assessment of costs and benefits of bird, animal, and plant control programs in the city.
h) Studies to determine if there are positive relationships between people and small urban animals, such as squirrels and pigeons, now serving as "community pets," that should be considered in planning.
i) Inclusion in the United States Census of questions on selected pets.
j) Attention to non-industrial, non-automobile related sources of air pollution in the design of urban buildings.

II-13. We recommend, for vacant urban lands, the use of naturally stable plant communities or similar horticultural alternatives to diminish maintenance costs without reducing benefits.

Research Recommendations

II-14. We recommend research on:
a) Appropriate sizes for manageable cities.
b) Criteria for ecologically sound population policies.
c) Criteria defining limiting factors and carrying capacities within specific urban areas.
d) The ramifications of megastructures, new towns, growth centers and experimental city development.

II-15. We recommend increased research on possible relationships between population density and human behavior.

II-16. We recommend research to determine whether increased diversity or increased living space will better ameliorate the problems of overcrowded areas.

II-17. We recommend research to allow creation of a broad, interdisciplinary model of migration. Various aspects of the urban system, i.e., economics, politics, climate, geography and population age-structure and density should be included. Such a model may increase the relevance and accuracy of interregional projections of migration.

II-18. We recommend further study of the motivations of those moving to and from urban centers as a means of predicting migratory patterns of ethnic and economic groups. Examination of majority attitudes towards social and economic minorities is also necessary.

II-19. We recommend research to develop a practical mechanism for translating the urban dweller's need for public outdoor green space into economic terms to compete with the need for housing or commercial development.

II-20. We recommend research on the effects that developments in information transport and the use of computers in communications will have on the urban way of life.

II-21. We recommend increased research on the ways transportation forms shape the city.

REFERENCES

Archibald, William S., and Stephen J. Kunitz, M.D. 1971. Detection of plague by testing serums of dogs on the Navajo reservation. HSMHA Health Report 86:377–380.

Beale, Calvin L. 1971. Rural-urban migration of blacks: past and future. American Journal of Agricultural Economics 53(2):302–307.

Beck, A. M. 1973. The ecology of stray dogs: a study of free-ranging urban animals. York Press, Baltimore.

Berry, B. J. L., ed. 1972. City classification handbook: methods and applications. Wiley-Interscience, New York.

Berzon, David R., Robert E. Farber, Joseph Gordon, and Elizabeth B. Kelley. 1972. Animal bites in a large city—a report on Baltimore, Maryland. American Journal of Public Health 62(3):422–426.

Bogue, D. J. 1969. Principles of demography. John Wiley and Sons, New York.

Boughey, A. S., J. B. Pick, and G. Schick. 1973. A migration model. Pages 397–419 in B. Dyke & J. W. MacCluer, eds. Computer simulation in human population studies. Academic Press, New York.

Boughey, A. S., R. P. Lowman, T. A. MacCalla, and J. J. Harris. An introduction to social ecology. Macmillan, New York. In press.

Brobst, D., H. M. Parrish, and R. B. Clark. 1959. The animal bite problem. Veterinary Medicine. May, 251–255.

Brown, H. J., et al. 1972. Empirical models of urban land use: suggestions on research objectives and organization. National Bureau of Economic Research, New York.

Bumpass, Larry, and Charles F. Westoff. 1970. The "perfect contraceptive" population. Science: 1177–1182.

Carding, A. H. 1969. The significance and dynamics of stray dog populations with special reference to U. K. and Japan. Journal of Small Animal Practice 10(7):419–446.

Cereghino, James J., M.D., Harold T. Osterud, M.D., Jacob L. Pinnas, M.D., and Monroe A. Holmes. 1970. Rabies: a rare disease but a serious pediatric problem. Pediatrics 45(5):839–844.

Clarkson, Frank E., Susan R. Vogel, Inge K. Broverman, Donald M. Broverman, and Paul S. Rosenkrantz. 1970. Family size and sex-role stereotypes. Science 167:390–392.

Cloud, Preston. 1971. Mineral resources in fact and fancy. Pages 71–88 in W. W. Murdoch, ed. Environmental resources, politics, and society. Sinauer Assoc., Stamford, Conn.

Coale, Ansley J. 1970. Man and his environment. Science 170:132–136.

Coale, Ansley J. 1972. The growth and structure of human populations. Princeton University Press, Princeton.

Committee on Intergovernmental Science Problems. 1972. Public technology: a tool for solving national problems. U.S. Government Printing Office, Washington, D.C.

Committee on Resources and Man. 1969. Resources and man. National Academy of Sciences-National Research Council, W. H. Freeman and Co., San Francisco.

Committee on Water. 1966. Alternatives in water management. National Academy of Sciences,

Washington, D.C.

Commoner, Barry. 1971. The closing circle. The Viking Press Inc., New York.

Davis, David. 1972. Rodent control strategy. Pages 157–171 in Pest control strategies for the future. National Academy of Sciences, Washington, D.C.

Deevey, F. S., Jr. 1960. The human population. Scientific American. Reprint, W. H. Freeman, San Francisco.

Doxiadis, C. A., Jr. 1968. Man's movement and his city. Science 162:326–334.

Doxiadis, C. A. 1970. Ekistics, the science of human settlements. Science 170:393–404.

Duncan, O. D., et al. 1960. Metropolis and region. The Johns Hopkins Press, Baltimore.

Esterlin, Richard A. 1961. The American baby boom in historical perspective. American Economics Review 51(5):869–911.

Ehrlich, Paul, and Jonathan Freedman. 1971. Population, crowding and human behaviour. New Scientist and Science Journal 50(745):10–14.

Eisenbud, Merril. 1970. Environmental protection in the city of New York. Science 170:706–712.

Emmons, Chester W. 1954. The significance of saprophytism in the epidemiology of the mycoses. N.Y. Academy of Science, Transactions, Ser 2, 17(2):157–166.

Emmons, Chester W. 1955. Saprophytic sources of *Cryptococcus neoxormus* associated with the pigeon (Columbia livia). American Journal of Hygiene 62:227–232.

Environmental Protection Agency. 1973. The quality of life concept. Washington, D.C.

Feldmann, Bruce M. 1974. The problem of urban dogs. Science 185:903.

Foa, Uriel G. 1971. Interpersonal and economic resources. Science 171:345–351.

Ford, Amasa B. 1970. Casualties of our time. Science 167:256–263.

Galarza, E., H. Gallegos and J. Samora. 1969. Mexican-Americans in the southwest. McNally and Loftin, Santa Barbara.

Galbraith, John Kenneth. 1964. Economics and the quality of life. Science 145:117–123.

Galle, Omer, Walter R. Gove, and J. Miller McPherson. 1972. Population density and pathology: what are the relations for man? Science 176:23–30.

Garrison, W. L. 1970. Computer technology, urban processes and problems, new futures for urban areas. World Future Society, Bulletin 3:1–16.

Goodwin, D. 1970. Feral pigeon; Columbia livia. Pages 54–60 in Pigeons and doves of the world. The British Museum, London.

Greenwood, Michael J., and Douglas Sweetwood. 1972. The determinants of migration between standard metropolitan statistical areas. Demography 9(4):665–681.

Hauser, Philip M. 1971. The census of 1970. Scientific American 225(1):17–25.

Hauser, Philip M., and Leo F. Schnore, eds. 1965. The study of urbanism. John Wiley, New York.

Hawthorne, V. M., and I. M. Lauder. 1962. Tuberculosis in man, dog, and cat. American Review of Respiratory Disease 85(6):858–869.

Hawthorne, V. M., W. F. H. Jarrett, I. M. Lauder, W. B. Martin, and G. B. S. Roberts. 1957. Tuberculosis in man, dog, and cat. British Medical Journal 2:675–678.

Holling, C. S., and Gordon Orians. 1971. Toward an urban ecology. Ecological Society of America, Bulletin 52(2):2–6.

Horsbrugh, P. 1972. Human-plant proximities: a psychological imperative. Indiana Nursery News 33(4). LaFayette, Indiana.

Hull, T. G., ed. 1963. Diseases transmitted from animals to man. 5th edition. Charles G. Thomas, Springfield, Illinois.

Human Settlement. 1971. Pages 9–16 in Sweden's national report to the United Nations on the human environment. Stockholm.

Iltis, H. 1972. Preliminary report on annotated bibliography on man's need for plants. Univ. of Wisconsin.

Istock, C. 1969. A corollary to the dismal theorem. Bioscience 19:1079–1080.

Jaffe, Frederick S. 1971. Toward the reduction of unwanted pregnancy. Science 174:119–127.

Jameson, D. L. 1970. The role of multidisciplinary and multiuniversity programs in today's multiversity. Proceedings of an invitational workshop on curriculum development in ecology and related environmental sciences. California State College Press, Los Angeles.

Keyfitz, N., and W. Flieger. 1971. Population: facts and methods of demography. W. H. Freeman, San Francisco.

Lampbear, F. O. 1971. Urban vegetation: values and stresses. Horticultural Science 6:332–334.

Land Use Subcommittee of the Advisory Committee to the Department of HUD. 1972. Urban growth and land development. National Academy of Sciences and National Academy of Engineering, Washington, D.C.

Lewis, C. A. 1973. People-plant interactions: a new horticultural perspective. American Horticulturalist 52(1), in press.

Lieberman, E. James. 1970. Reserving a womb: case for the small family. American Journal of Public Health 60(1):87–92.

Louder, D. R. 1971. Migration and population trends in the Pacific northwest. University of Washington Business Review 23–44.

McGinnis, Robert. 1968. A stochastic model of social mobility. American Sociological Review 33:712–722.

McHarg, Ian. 1969. Design with nature. Natural History Press, Garden City, New Jersey.

McPherson, M. B. 1971. Management problems in metropolitan water resources operations. Technical Memorandum No. 14, Urban Water Resources Research Program, ASCE, New York.

Malin, H. Martin. 1973. Toward a national energy policy. Environmental Science and Technology 7(5):392–397.

Mandelker, D. R. 1971. Managing our urban environment. Contemporary Legal Education Series, Bobbs-Merrill.

Martin, Russell J., Paul R. Schnurrenberger, and Norman J. Rose. 1969. Epidemiology of rabies vaccinations of persons in Illinois, 1967–1968. Public Health Report 84(2):1069–1077.

Marx, Leo. 1970. American institutions and ecological ideals. Science 170:945–952.

Marx, Martin B., and Michael L. Furcolow. 1969. What is the dog population? Archives of Environmental Health 19:217–219.

Meyer, J. R., J. F. Kain, and M. Wohl. 1965. The urban transportation problem. Harvard University Press, Cambridge, Mass.

Meyer, K. F. 1959. Some general remarks and new observations on psittacosis and ornithosis. WHO/FAO Expert Committee on Zoonoses.

Michelson, W. 1970. Man and his urban environment: a sociological approach. Addison-Wesley, Reading, Mass.

Milgram, Stanley. 1970. The experience of living in cities. Science 167:1461–1468.

Morrison, P. A. 1969. Duration of residence and prospective migration: the evaluation of a stochastic model. Demography 6:553–561.

Morrison, Peter A. 1971. Chronic movers and the future redistribution of population: a longitudinal analysis. Demography 8(2):171–184.

O'Keef, P. 1972. Pet food institute fact sheet, 1972–1973. Pet Food Institute, Chicago, Illinois.

Packard, Vance O. 1972. A nation of strangers. McKay, New York.

Parrish, Henry M., Frank B. Clack, Duane Brobst, and James F. Mock. 1959. Epidemiology of dog bites. Public Health Report 74:891–903.

Peterson, James T. 1969. The climate of cities: a survey of recent literature. National Air Pollution Control Administration Publication #AP 59, Durham, North Carolina.

Pick, J. B. Pyramid—a graphic population program. Demography. 1974 (in press).

Pickhard, J. 1971. United States metropolitan growth and expansion, 1970–2000. Population projections. Urban Land Institute, Washington, D.C.

President's Task Force on Science Policy. 1970. Science and technology: tools for progress. U.S. Government Printing Office, Washington, D.C.

Quarterman, K. D., W. C. Baker, and J. A. Jensen. 1949. The importance of sanitation in municipal fly control. American Journal of Tropical Medicine 29:973–982.

Reif, J. S., and D. Cohen. 1970. Canine pulmonary diseases and urban environment. II. Retrospective radiographic analysis of pulmonary diseases in rural and urban dogs. Archives of Environmental Health 20(6):684–689.

Rogers, A. 1968. Matrix analysis of interregional population growth and distribution. Univ. of Calif. Press, Berkeley and Los Angeles.

Royal Ministry for Foreign Affairs and Royal Ministry of Agriculture. 1972. Urbanization and planning in Sweden. Stockholm.

Russell, C. J., and W. O. Spofford, Jr. 1972. A quantitative framework for residuals management decisions. In A. V. Kneese and B. T. Bower, eds. Environmental quality analysis: theory and method in the social sciences. The Johns Hopkins Press, Baltimore.

Sears, Paul B. 1958. The inexorable problem of space. Science 127:9–16.

Schmid, J. 1973. The environmental impact of urbanization. M. S. in preparation, Department of Biological Sciences, Bernard College, New York City.

Schnurrenberger, P. R., R. J. Martin, G. L. Meerdink, and N. J. Rose. 1969. Epidemiology of human exposure to rabid animals in Illinois. Public Health Report 84(12):1078–1084.

Schwirian, Kent P., and Anthony J. LaGreca. 1971. An ecological analysis of urban mortality rates. Social Science Quarterly 52(3):574–587.

Smith, F. 1972. Man in the urban environment. Rockefeller Foundation, New York.

Smith, J. 1972. The Gateways: parks for whom? Pages 213–236 in National parks for the future. The Conservation Foundation, Washington, D.C.

Solow, Robert M. 1971. The economist's approach to pollution and its control. Science 173:489–503.

Stein, R. G. 1971. Architecture and energy. Preprint, AAS Philadelphia Meeting, December 1971.

Stouffer, Samuel A. 1960. Intervening opportunities and competing migrants. Journal of Regional Science 2(1):1–26.

Tabah, L. 1968. Representation matricielle des perspectives de population active. Population 3:437–476.

Taeuber, Conrad. 1972. Population trends in the 1960's. Science 176:773–777.

Tansey, M. F., and R. P. Roth. 1970. Pigeons: a new role in air pollution. Journal of Air Pollution Control Association 20(5):307–309.

Thompson, W. R. 1965. A preface to urban economics. The Johns Hopkins Press, Baltimore.

Van Arsdol, Maurice D., and Leo A. Schuerman. 1971. Redistribution and assimilation of ethnic populations: the Los Angeles case. Demography 8(4):459–480.

Ward, B. 1972. An urban planet? Architectural Forum 137(5):30–39.

Webb, C. H. 1965. Pets, parasites, and pediatrics. Pediatrics 36:521–522.

Wingo, Lowdon, Jr. 1963. The use of urban land: past, present, and future. Resources for the Future, Inc., Washington, D.C.

Woldow, N. L. 1972. Studies in the reproductive capacities of birds with a field study of the feral rock dove, Columba livia, in the urban environment. Doctoral dissertation, Rice University (University Microfilms).

Wolff, A. H., N. D. Henderson, and G. L. McCallum. 1948. Salmonella from dogs and the possible relationship to Salmonellosis in man. American Journal of Public Health 38(2):403–408.

Wrigley, E. A. 1969. Population and history. McGraw-Hill, New York.

SECTION THREE:
Processes:
Interventions, Indicators, and Models

To assess the condition of the urban ecosystem adequately and to respond appropriately, we need to have reliable measurements of the state of the system and to know the consequences of our actions. We should be able to adjust the relationships between urban structure and function and to intervene wisely in urban resource flows. Our discussion here concerns the kinds of interventions that appear possible, the criteria for indicators of urban system health, and the models useful in understanding the system.

When we speak of intervention, we are not using the word in its usual senses, i.e., to enter as an irrelevant or extraneous feature or circumstance, or to fall between points of time or events. Rather, as we use the term, intervention is rational, well-considered action taken to alter aspects of urban system performance. Decisions to intervene in the urban system should be based on comprehensive knowledge of the whole urban system and its natural support systems, on the nature of the system's components and their interrelationships, and on the particular characteristics of the problem being addressed. Intervention is not the province of government only, nor of other institutions charged with responsibility for aspects of the urban system. Decisions to intervene, if they are to be wise decisions, will consider the ecological, socio-psychological, economic, technological and political dimensions involved. Consideration should also include urban residents' perceptions of city problems, the capacity of the natural environment to accept modifications, the ecological costs of intervening (or not intervening), and the political and technological feasibility of proposed action. Intervening in the urban system, therefore, is not a matter of serendipity; rather, it is a matter of wise decision-making based on comprehensive understanding of all the information on system processes available. Our discussion here is intended to help provide the basis for wise interventions.

Several federal laws already affect the physical processes of the urban system. The most significant, perhaps, is the National Environmental Policy Act (NEPA) of 1969 (Public Law 91–190). It directs all federal agencies, to the fullest extent possible, to orient their policies, plans and programs to: meet national environmental goals; encourage productive and enjoyable harmony between man and his environment; promote efforts preventing or eliminating damage to the environment and biosphere; stimulate the health and welfare of man; and enrich the understanding of the ecological systems and natural resources important to the nation.

Section 102(2) (C) of the Act and the guidelines prepared by the Council on Environmental Quality require federal agencies to consider carefully, during decision-making, the environmental aspects of proposed action. Environmental impact studies must proceed concurrently with preliminary technical and economic studies. Further, federal agencies are required to: circulate their draft environmental statements to other agencies—federal, state and local; make the statements available to the public; weigh the responses of other agencies and the public; and prepare final environmental impact statements that are responsive to the comment received.

The purposes of the environmental assessment, comprehensive consultation and review procedures are: to provide agencies an opportunity to understand the potential environmental effects of proposed actions; to avoid adverse effects; to restore or enhance environmental quality; to explore alternative actions that will avoid or minimize adverse actions; to evaluate both short- and long-term effects of proposed actions on man, his physical and social surroundings and on nature; and to balance the results of environmental assessments against the net economic, technical and other benefits.

In addition to NEPA, the amended Clean Air Act and the National Water Pollution Control Act of 1972 (PL 92–500) exert environmental controls on air and water quality standards. Any action of a federal agency and any federally-funded action are subject to the requirements of the National Environmental Policy Act. Within urban ecosystems, the planning, design, implementation and operation of transportation systems, sewage systems, flood control projects, water supply projects, federal buildings and other major actions are covered by the Act. The process of the environmental impact statement assures full disclosure of federal planning. It also encourages participation of federal, state and local agencies and citizen groups.

The Water Pollution Control Act requires any person, firm or public agency discharging pollutants into navigable streams to obtain a permit. The discharges must meet effluent standards imposed by the responsible state agency and the Environmental Protection Agency (EPA) within the period specified in the Act.

These laws, regulations and guidelines, in harmony with international goals and agreements, are affecting many institutions concerned with urban ecosystems. Some of the problems associated with these regulations may be resolved by studies that will: define "major action", or rank actions according to degree of impact on the environment; upgrade the "state of the art" of environmental assessment procedures and the analysis of short- and long-range environmental impacts; and assess the technical, economic and ecological ramifications of applying NEPA requirements to the private sector of society.

ALTERING RELATIONSHIPS BETWEEN URBAN STRUCTURE AND FUNCTION

The physical processes of the urban ecosystem involve energy flow, materials cycling and information transfer. Information transfer is accomplished through energy flows, material movements, and behavioral responses. In urban ecosystems, as in all systems, energy is characterized by an ordered input and a disordered output. In the past, technology has focused on input. If we are to conserve materials and energy, we must develop an output technology as well. Improved technology can manipulate energy dissipation through the structural characteristics of configuration, differentiation and density.

Material cycling in the urban system presents more fundamental problems. Characteristically, the natural systems dependent on imported energy and materials selectively concentrate materials. The greater the ratio of imported energy to total energy, the greater the concentration factor. Urban ecosystems, largely dependent upon imported energy, tend to concentrate materials, products and pollutants.

It is not clear to what extent human environmental conditions may be improved by manipulation of urban physical structure and function. Certainly, we need a greater understanding of the effects of scale on energy flow, materials cycling and structural parameters, and of the nature and scale of the unintended distribution of materials by various transport systems.

Attempts to Improve Urban Ecosystems

In attempting to improve the relationship between the physical structure and function of urban systems, we need to identify appropriate alternatives for improvement, those which, if presented and specified effectively, could be put into action through the urban decision-making process. Identifying and evaluating courses of action depend upon comprehensive knowledge of the urban and natural systems involved. Improvement of our knowledge base is fundamental.

We may describe an urban system as being in a particular state at a particular time. It arrives at that state as a result of previous states and of the forces altering them.

Both controlled and uncontrolled elements affect change in the urban ecosystem. A controlled element might be the supply of land determined by zoning ordinances. The uncontrolled elements, the state of the system, and the particular point in time are vectors. Residential density at a particular time is an example of the state of the system. Average temperature or land costs are uncontrolled elements.

The urban planner, policy maker or citizen intent on improving the system's performance bases his actions on his knowledge of how the system works. He recognizes that without intervention, a set of states or a path will occur. The state of the system in the past affects its present state, in turn affecting the state of the system in the future. Interventions, however, can create alternative pathways, thus affecting the state of the system now and its state in the future. Sometime in the future, we can again intervene to change the state of the system.

Intervention will improve performance of

the urban ecosystem if the alternative path chosen is more desirable than the non-intervention path.

Appropriate intervention is complicated by the question "what is more desirable?" A first test of desirability is whether or not a course of action is consistent with emerging human goals and aspirations. If it is, social change mechanisms may be used to select the desired alternatives. Alternatives, however, are seldom clear-cut. Urban land has many potential uses, e.g. gardens, parking lots, parks or public housing. If meaningful choices are to be made, it seems necessary to develop methods which will describe and evaluate alternatives more explicitly.

Opportunities for Improvement

In examining urban ecosystem performance, we see three classes of opportunities for improvement. We may, first of all, bring uncontrolled elements of the urban system under control. We could, for instance, control sulphur dioxide emissions by taxing or fining those responsible for the emissions. It is unclear, however, to what extent this class of intervention provides opportunities for improvement since many control mechanisms are not inherent in urban ecosystems as they are in natural systems. Some controls, exercised at the national level, are only gross instruments when applied at the urban level.

Secondly, we may intervene by changing the nature of the relationships involved—to create an alternative course of action by changing the way the system operates. New styles of living could be adopted, for instance. Historically technological development has been the most powerful force behind such changes and may allow us to change the way a system operates.

Thirdly, we may intervene in the process by which alternatives are selected. We could, for example, increase the decision-maker's knowledge of how facets of the urban system interrelate and of the system's ecological dimensions.

Presumably, an informed decision-maker will make wiser decisions.

Constraints on Improvement

An involved community, as indicated in an earlier part of the book, should educate itself on environmental and biological matters, set ecologically sound goals for the urban ecosystem, and implement them. The choice of alternatives open to such a community should be consistent with its aspirations; these alternatives constitute one set of constraints on improvement. Other constraints are supply limitations, stress limits, available technology and the like, which restrict the ways resources are used to achieve specified goals. A particular time and its technology impose limits on human functioning and on the functioning of urban ecosystems.

Before elevators were developed, for example, the heights of buildings in the city were limited; this limitation affected density and living space relationships. Today we are finding that the elevator is a mixed blessing. The elevator is essential for transport in tall buildings. However, in some situations, such as apartment complexes, the use of tall buildings, made possible by elevator technology, affects social relationships, e.g., parents may be reluctant to allow their children to play at ground level while the parents remain on the twenty-third floor.

The supply of available resources poses another constraint on attempts to improve the urban system. Throughout history, supply limitations have been circumvented by new technology or by substitution of readily accessible materials. Finding alternatives to those resources in short supply may improve urban system performance, though the search for new resources often consumes large amounts of energy. Patterns of resource use can be changed when such changes are consistent with the goals of the community and will lead to a desired alternative state.

Man himself imposes arbitrary constraints upon his functioning and the functioning of other elements in the urban system. The current concern for air pollution is an example. Of many air pollutants, we know only that too much of them will kill, injure or limit organisms in the ecosystem and that small amounts appear relatively harmless. Consequently, stimulated by a high level of pollution, the political process sets allowable limits for pollutants, but without accurate knowledge of all the factors involved. Political mechanisms are also used to set water quality standards, to limit the use of pesticides and to impose other constraints.

Clearly we need to gain reliable information about tolerable limits for dangerous elements. New knowledge should be communicated to those involved in the political process, to enlarge their capacity for setting meaningful limits.

Despite the various constraints, there are opportunities for improving urban ecosystems. The following specific suggestions, particularly those for incorporating the natural environment into urban ecosystems, indicate what may be done.

Man's interaction with nature can be improved through careful physical shaping of urban areas. Proper guidance of private and public development will keep land use compatible with ecosystem processes and natural resources. The implementation of new technology can help maintain or even improve the natural environment, overcome resource limitations or improve resource utilization. Frequently, the implementation of technology has, instead, substituted man-made attributes for natural ones, at high economic and environmental costs. In our use of new urban ecosystem technologies, natural ecosystem elements should be utilized wherever possible; new technology should complement, rather than replace, existing natural elements. Improved organization of urban areas can reduce conflict between the natural and man-made environment. Environments reflecting planned diversity and improved connectivity at suitable scale can increase community stability and provide individual satisfaction.

Traditionally, open spaces have been used primarily to provide recreational resources, to act as buffer zones between adjacent incompatible land uses, and to add "design" to urban form. Open spaces have other functions, however. Numerous animal and plant species are supported by woodlands, particularly along the edges of open spaces. The cooling effect of open space alters adjacent microclimates. Open green space improves urban air quality. It can also function as an educational resource.

The economic short-term benefits of land development are obvious and often impede successful acquisition and preservation of open space in a rapidly urbanizing region. When the full range of natural resource, land use and human benefits of open space are considered, its acquisition becomes more easily justified.

The "dead" land associated with modern highway systems—highway embankments, interchanges and dividers—are commonly used, beyond their highway safety function, only to grow grass and to hold unstable slopes. The use of natural vegetation, for example prairie or shrub cover, may substantially reduce maintenance costs. With improved design and management, such areas could be aesthetically satisfying and fulfill their present functions at lower costs. These areas may also provide storm water storage, reducing the size and costs of community storm sewers, and retain silt and nutrients to create rich patches of land for agriculture.

If opportunities for multi-purpose use of urban open land are pursued, future urban configurations may improve urban ecosystem performance. We need to identify known opportunities and constraints and to correlate them to preferred or potential land uses.

Early in the identification of these opportunities, possibilities for natural environmental management should be presented to urban decision-makers in forms compatible with current land use planning approaches, classification procedures and control measures.

At later stages, possible approaches to opportunities for improving the urban system will require modification of local, national and international environmental policies and planning, as well as more knowledge of urban ecosystems. Planning should emphasize the interdependence between the physical structure of the city, land use patterns, social behavior, energy utilization and the biological nature of urban ecosystems.

INTERVENING IN THE MATERIALS AND ENERGY SYSTEM

If we are to reduce undesirable effects of resource extraction and residuals discharge, we must change the nature and size of flows into, through, and out of the materials and energy segment of the urban system. We attempt here to classify methods for altering materials and energy flows in our efforts to reduce resource depletion or residuals discharge. Hopefully, our discussion will help to identify and to overcome the obstacles hindering effective implementation of strategies leading to such reductions.

Our discussion here centers on man as "economic man" or "man-the-consumer". Man is, however, more than an economic man: he has spiritual, cultural, psychological and physiological dimensions as well. Thus, although we are primarily concerned with the effects that particular interventions may have on man-as-consumer, we must bear in mind that changes in energy and material flows can affect, and can be affected by, the whole man, with all his dimensions.

We also discuss here technological changes which may improve the urban flow system. The capacity to initiate technological change enables man to alter the patterns of both resource use and residuals discharge. However, we must be careful not to assume that "tinkering" or "fine-tuning" the flow system will solve all its problems. If the urban flow system is on the verge of collapse, as it sometimes appears to be, technological "tinkering" is not a realistic solution. In such cases, our values and life styles may have to change and our patterns of resource use and residuals discharge will be seriously altered. It will be necessary for man, the spiritual, cultural, psychological and physiological being who understands ecological principles, to restructure the urban flow system to meet his many needs.

Since the urban population demands goods, services and energy, resource depletion and residuals discharge are urban problems. Solutions to perturbations in resource flows within the city may lie outside it. An urban physical support system lies only partly within a particular urban administrative unit; the affected environmental system may lie entirely outside the city. Some of the interventions we identify and some of the measures necessary to implement them may lie entirely outside urban boundaries and outside urban jurisdiction as well; these same interventions may affect only the urban ecosystem. Other actions may be taken in the city and have effects far beyond its boundaries. Our discussion of interventions into the materials and energy segment of the urban system, then, is not limited to purely "urban" problems or "urban" situations, but to those problems and solutions relevant to the urban situation.

As the rate of resource depletion or residuals discharge increases in response to increased consumption, disutility increases. By "disutility" we mean the options that are foregone, in terms of present or future possible uses or enjoyment of resources or of environments that are adversely affected by resource withdrawal or residuals discharge.

With renewable resources, such as hydroelectric power, or with the assimilative capacity of the environment, including waste degradation, the disutility associated with resource depletion or residuals discharge depends directly on: 1) the rate of resource extraction or residuals discharge; 2) the lost use or satisfaction per unit of extracted resource or discharged residual; and 3) the fraction of the total capacity of the environment (to yield resources or to assimilate residuals) that is exploited. (See Fig. 1).

If the above relationship is written in terms of the total stock of a productive resource or the total capacity of an environmental sink, the relationship can apply to nonrenewable

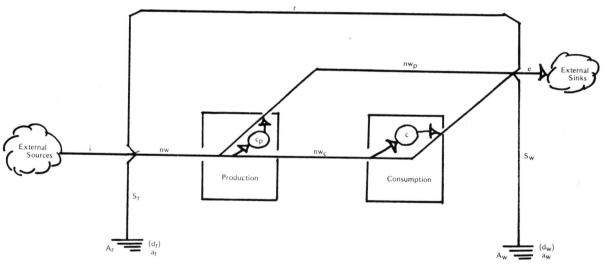

A_r — stock of nonrenewable resources
A_w — capacity to accept residuals
a_r — capacity of environment to yield renewable resources
a_w — capacity to assimilate biodegradable residuals
S_r — rate of resource extraction
S_w — rate of residuals discharge
d_r — disutility per unit of resources extracted
d_w — disutility per unit of residuals discharged
i — rate of import of resources
e — rate of exports of residuals
n — rate of product consumption
w — amount of resource input per unit of product consumed
w_p — amount wasted from production

w_c — amount wasted from consumption
c — stock of consumer capital
c_p — stock of productive capital

For renewable resources or assimilative capacity:
$$D \alpha d \times S \times \frac{S}{a} \alpha \frac{dS^2}{a}$$
For the total stock of a productive resource or the total capacity of an environmental sink:
$$D \alpha d \times S \frac{1}{\overline{\frac{A}{S}}} \alpha \frac{dS^2}{a}$$

Figure 1   Mass Balance Design

resources (e.g., fossil fuels) or to the finite capacity of the environment to accept refractory residuals (e.g., a land-fill site).

We assume, in our discussion, that increasing consumption implies corresponding increases in resource extraction or residuals discharge; the rate of extraction or of discharge varies directly with the rate of product consumption. The total environmental disutility of a product may be reduced in two ways. First, the disutility associated with resource extraction or residuals discharge can be decreased, e.g., by treatment of a consumption residual before its discharge. Second, the total stock of nonrenewable resources and the total assimilative capacity of the environment or the capacity of the environment to yield renewable resources and assimilate biodegradable residuals can be increased (e.g., by discovering previously unknown resources or by in-stream aeration to increase the assimilative capacity of a river). Depletion of resources or saturation of assimilative capacity inevitably occurs with time; total disutility then decreases as a result of accumulative depletion or saturation.

Resource use and residuals discharge can be manipulated in several ways. They may be reduced without lowering consumption levels. In such situations, a reduced rate of resource extraction or of residuals discharge can bring the optimal consumption rate (i.e., that rate which balances environmental and market considerations) closer to the actual market rate and increase the total utility of a product (i.e., the satisfaction derived from both the

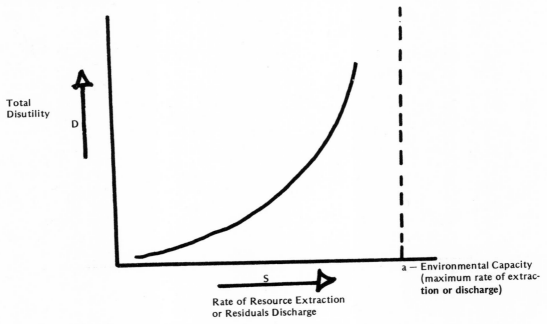

$D \propto \frac{dS^2}{a}$, where d = disutility per unit of resources or residuals flow.

Figure 2a

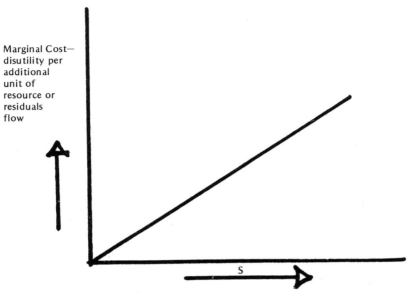

$\frac{\delta D}{\delta S}$ or the slope of the curve in Figure 2a. Disutility per unit of resource or residuals flow increases as rate of resource extraction or residuals discharge rises.

Figure 2b

product and the environment). Resource and residuals management can also involve changes in consumption patterns. A change in the quality of a product (by substitution of a less critical resource, if it reduces the desirability of the product) may reduce the satisfaction consumers derive from the product. If money usually spent on a product is instead expended to control the pollution associated with that product, the product will cost more on the market, partially offsetting the increased consumer satisfaction that may result from improvement of the environment. If consumers of a product determine to value acquisition of additional items less, fewer items will be consumed and the product will be consumed at a level more fully representing a balance between desired environmental quality and product usefulness.

We have not yet considered the effects of changes in population size on resource use and residuals discharge. For example, if the utility of a product, per capita, does not change although the size of the population doubles, the given level of marginal utility (satisfaction with the product) will occur at twice the original consumption level; if the market cost of a product remains constant, consumption of the product will double. If consumption doubles, however, the difference between the total marginal cost of the product and its market cost will increase, due to the increased adverse effects on the environment; the difference between the actual consumption level and the optimal level will increase, decreasing total satisfaction proportionally more than the increase in population.

One environmental cost associated with increased resource and residuals flows which should be noted is the concomitant increase in the amount of space devoted to resource extraction and residual discharge activities. Because the demands for space associated with material and energy flows compete with the demands of man's increasing stock of capital goods, it is conservative to assume that the total disutility (total adverse effect) associated with a product decreases linearly with the fraction of the production or amenity resource that is utilized.

Classes of Interventions

The places at which man can intervene in the materials and energy flow system to reduce the disutility stemming from resource depletion or residuals discharge fall into the seven general categories that follow.

1) **Increasing Environmental Capacity.** Environmental capacity includes the capacity to yield renewable or non-renewable resources and the capacity to accept degradable or refractory residuals. Resource yields can be increased, for instance, by: impounding streams to increase hydroelectric power output; seeding clouds for rain; developing processes for using solar energy or low grade ores; or discovering a previously unknown mineral or energy source. Interventions improving the assimilative capacity of the environment include stream aeration or flow regulation to enlarge the stream's capacity to accept effluents or the use of improved land-fill techniques to allow larger quantities of solid wastes to be put into smaller areas.

2) **Decreasing Disutility per Unit of Resource or Residual Flow.** The disutility per unit of resource extracted or of residual discharge can be reduced by altering the distribution (in time and/or space) of extraction or discharge patterns, or by changing the composition of residuals. Actions controlling distribution in time include seasonal restrictions on fish or game harvest and controlled rates of pollutant discharge compatible with the environment's assimilative capacity. Actions controlling spatial distribution include the extraction of resources at locations where the environmental effects are less significant, effluent discharge in places with a greater capacity to accept them, and effluent dilution or dissipation to avoid local concentrations. Treating wastes at the point of generation (for instance, at an apartment incinerator) in contrast to a central facility (public incinerator or land-fill) affects spatial distribution as well.

The form or composition of residuals can be altered to decrease adverse effects. Toxic elements can be captured and concentrated. Components like water or air can be separated from residuals and returned to the environment with no ill effects. Or residuals can be degraded to the less complex forms more easily assimilated by the natural environment. For example, air pollution may be controlled in part by precipitators or other devices reducing stack emissions. Automobile emission controls, sewage and industrial waste treatment, shredding or baling solid wastes, and "permanent" packaging of toxic or radioactive wastes prior to disposal are other examples.

3) **Importing Resources and Exporting Residuals.** Because resources are not uniformly distributed over the surface of the earth, every nation depends on imports. An urban system ordinarily provides no resources except possibly water or mineral aggregates. The system thus depends on imports and, additionally, requires land beyond its boundaries. Odum (1970) estimates that to support one person in a healthy environment in Georgia's climate requires five acres of land, or about 100 times the average per capita space pro-

vided in an urban area. Most of the space is needed for the production of food and fiber and for watersheds, waste disposal sites and the like.

Wolman (1965) presents this summary of selected inputs and outputs of a hypothetical U.S. urban community of one million persons:

Inputs (tons/day)	Outputs (tons/day)
Water—625,000	Waste water—500,000
Food—2,000	Solid waste—2,000
Fuel—9,500	Air Pollutants—950

In a conceptual sense, an urban system extends from the source of its resources to the point of affluent disposal. There is, in this sense, no importing or exporting. The reality of political boundaries, however, makes importing resources and exporting residuals a fact of urban life.

4) **Reclamation.** In our discussion of reclamation, we blur the distinction between reclamation and recycling. As Spofford describes it, reclamation "... refers to residuals from production activities which become inputs into a production process, either the same one in which they were generated, or another process—which can be at the same or a different location." He states further, "The factor (criterion) which distinguishes recycling from reclamation is that the recycled residual has not gone through the hands of a 'user'; it is a production residual or an intermediate product, not a 'final product' or a residual from a consumption activity" (Spofford, 1971).

For our purposes, production includes not only industry but all processes generating goods and services. Some production residuals are indistinguishable from consumption residuals because they are identical (scrap metal is identical whether discarded by government, industry, commerce or an individual), or because they are mixed in public sewage or solid waste collection systems.

Waste paper, glass, or metals may be refabricated and reused. Whole bottles may be refilled and reused. Energy can be recovered by the combustion of solid wastes or waste water solids. Low grade wastes can be utilized for direct heating or desalination. In buildings, heat from lighting fixtures may contribute to space heating. Nutrients can be recovered by processing sewage sludge to yield fertilizer or by spray irrigation of effluents. Land-fill disposal has been suggested as a means of storing resources for further use.

Water, the most abundant resource in the urban system, can be reclaimed as well. Industrial reuse is already widely practiced; direct recycling of municipal waste waters is being approached cautiously. Artificial recharging of ground water supplies is another possible reuse for waste-waters.

5) **Production Changes.** Production includes the extraction, generation, manufacture, transportation and delivery of goods, services, electricity, fuel, food and water by government, commerce and industry. Industry includes North American agricultural operations which depend almost entirely on inputs of fossil fuel energy, goods and services from other production segments, (in contrast to ancient agricultural operations which relied more immediately on solar energy).

Intervention by production changes includes those changes that can alter resource inputs and residuals outputs without affecting the nature and magnitude of the product output. For example, the amount or composition of production capital, such as buildings, public work structures, transportation or assembly line equipment, can be altered without affecting the products. The material content of the capital stock, the resource requirements and residuals discharges necessary for its production, or the electricity or water required for its operation may be reduced. Transportation and physical facilities account for the largest share of production capital. Transportation changes may stem from changes in urban design, transportation systems or vehicle design. Reducing material inputs and energy consumption in industrial buildings can be accomplished by changing building designs, heating and lighting systems, or siting.

Recycling can decrease both resource demands and residuals discharges by keeping materials in the production stream. Industrial recycling of water and some metals offer impressive examples.

6) Altering Product Specifications. Product specifications can be changed to alter the composition or weight of products to reduce the amounts of environmentally sensitive resources they contain or the nature of the residuals they will yield when discarded. Some changes will not affect the number of items consumed, but product changes often do alter consumer preferences and thus affect consumption levels. Specifically, any product that performs more efficiently, e.g., by providing better service, by being more durable, or by consuming electricity, water or fuel more efficiently, may reduce the need to replace the product or the amount of electricity, water or fuel the product uses. In such cases reduction of resource demands or residuals outputs will reduce adverse environmental affects; consumers will gain increased satisfaction from the product. These effects combine to increase total well-being by bringing the market level of consumption, and the "optimum" level closer together (see Fig. 1).

Producing less elaborate packaging, or non-phosphate detergents to reduce water pollution, are ways to alter the specifications of disposable products to reduce their content of environmentally sensitive materials. Changes may also be made to reduce resource input or residuals discharge per unit of product output. For example, steel can be substituted for aluminum because steel manufacturing requires less energy. Product specifications can also determine the ease with which materials can be reclaimed, either for reuse (as with bottles) or refabrication (as with cans).

Specifications for capital goods can also be used to control resources and residuals: by limiting the content of critical materials in products; by substituting for or eliminating non-functional components; by reducing amounts of critical resources used or residuals generated in production; by making goods easily reclaimable, durable, and easy to maintain, repair and update; by reducing residuals discharged by the product; and by making the product more efficient.

7) Reducing Consumption. If products remain unchanged, reduced consumption rates require either changes in the public's attitude toward products and consumption, or changes in the price structure. If either or both changes occur, consumption will be reduced by reduction of the rates at which personal capital is acquired or by reduction of the consumption of disposable products.

Acquisition of capital goods at a reduced rate may reduce resource demands and residuals outputs by reducing the amount of materials in the stream of goods entering the production sector, by reducing the energy and materials flow inherent in production of personal capital, and by reducing the flow of fuel, water or electricity associated with the operation of capital items.

Capital acquisition rates can be reduced by decreasing the rate at which capital stock is discarded, or by reducing the total amount of stock. Existing capital stock is discarded at a rate dependent both on product lifetime and on owner attitude. Product lifetime depends in part on product specifications. Owner attitudes depend on whether he is satisfied to use a worn or obsolete product or one he has grown tired of; his attitudes also depend in part on whether or not a replacement is available. Reducing capital stock, or at least a reduced rate of increase of capital stock, implies willingness to forego some product utility by accepting less, by doing without, by use of public or corporate services as a substitute for services previously derived from private capital (e.g., the use of public transportation systems or laundromats in place of private automobiles and washing machines), by the shared use of private capital (e.g., car pools or communal households), etc.

Reduced consumption of disposable products may result from willingness to forego unnecessary or frivolous goods or services or from specifications improving the efficiency of capital products that use fuel, water or electricity. Reduced use of these products can also result from more prudent use of equipment or appliances (e.g., lowering furnace thermostats at night, when one is absent from the house, or turning off unnecessary lights).

EVALUATING ALTERNATIVE INTERVENTIONS

Not all interventions affect the material and energy flow system. Increasing the capacity of the environment to yield resources or assimilate wastes, for example, leaves the system untouched; it also fails to address the basic questions of resource depletion or residual discharge. Increasing environmental capacity requires that more nonrenewable resources be discovered, that new places to hide refractory wastes be found, or that the rates at which renewable resources can be extracted or wastes assimilated be increased. As a short-

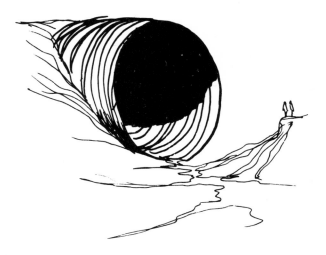

term approach to resource and residuals problems, this approach can be useful or even necessary. However we cannot rely indefinitely upon the environment, and our technological capability to exploit it, to accommodate ever increasing rates of resource extraction and residuals discharge.

The importing of resources and the exporting of wastes, likewise, does not interfere with the materials and energy flow system. Such action reduces environmental costs merely by externalizing the environmental effects that are not reflected in the market cost of resources or waste disposal. Importing materials into a densely populated urban area is necessary; conservation measures may reduce this need, but they cannot eliminate it entirely. Likewise, exporting wastes is an unsuitable answer to urban residuals problems. Many urban areas have encountered difficulties in obtaining land outside their boundaries for land-fill sites. Similar problems have been encountered with sites for the spray disposal of effluents or the disposal of liquid wastes.

Storage or treatment of wastes to reduce their adverse effects, an obvious and common solution, has the advantage of attacking a particular problem directly, but it has two specific drawbacks: first, the law of diminishing return applies; succeeding steps in residuals improvement become more difficult and expensive to achieve. Second, the potential resource values in residuals are ignored by storage or treatment. Complex organic structures are broken down and material concentrations are dissipated, with concomitant dissipation of energy and resources.

Production changes, product specification changes, reduced consumption, reclamation, and direct control of extraction or discharge rates are interventions which affect resources and residuals simultaneously. For example, if gasoline is rationed in California to reduce air pollution, fuel resource consumption is also affected. If smaller, less powerful, fewer or more durable automobiles are produced, resources will be conserved and residuals reduced. The value of coal as a substitute for gas and oil is diminished by those characteristics which make it a source of air pollution.

Despite these examples, we have no guarantee that every change which reduces demand will improve the residuals situation or vice versa. Product or manufacturing changes to produce less damaging residuals may require the input of critical resources. Similarly, changes that reduce demands on a critical resource may not change, or may even have an adverse effect on, waste disposal. The breeder reactor, for example, will conserve nuclear fuel stocks, but disposal of effluents may be a critical problem.

Intervention into the energy and materials flow system may involve tradeoffs of one resource against another or against residuals (pollution) effects. Adopting air pollution control devices for automobiles, for example, may involve a 15 per cent to 30 per cent decrease in efficiency (Hammond, 1972; Lincoln, 1973). Reclaiming residuals involves energy input and possibly the creation of new waste problems. Longer-lived products may have higher resource demands for their operation and maintenance. Accordingly, proposals are often made to demolish old buildings or to replace inefficient power plants. Such proposals should be weighed, however, against the fact that preservation and restoration of old buildings conserve the materials and energy which would be required to replace them. Further, old buildings are frequently more energy efficient than the new buildings replacing them.

The Indirect Costs of Intervention

The effects of any intervention in the materials and energy flow system cannot be measured entirely in terms of resource depletion and environmental degradation. More efficient resource extraction or the substitution of one resource to conserve another may reduce environmental amenity resources. Increased reliance on coal means increased stripmining. Development of a breeder reactor, which can use lower grade ore, may require the excavation and processing of large quantities of granite.

"Present reactors do not require too much land disturbance, but breeder reactors would cause a very great deal. It has been estimated that fifteen million tons of granite would need to be mined and processed each day, if the world's energy economy of the future is based on breeder reactors; this is about twice the present rate of coal mining in the world" (Cook, 1972).

An intervention altering the rate or pattern of a production activity will have economic repercussions and may produce political reactions. Diverting industrial production from one item to another will require economic adjustments that can result in local unemployment and personal hardships, even though the long-term benefits may be significant. Examples abound. Pollution control requirements have forced marginally efficient industries to close; restrictions on resource extraction have forced industries such as fisheries to stop or restrict their activities. The challenges of air pollution, energy conservation and other problems associated with an automobile-oriented society may encourage the development of an alternative to the automobile. If one is developed, the economic disruption resulting from reduced purchases of the automobile will affect a major segment of the American economy.

Importing resources on a national scale involves the indirect political and economic costs that are associated with dependence on foreign governments for continued reliable supplies; the balance of payments deficit is a further complication.

Moral implications are involved as well: six per cent of the world's population consumes one-third of the world's nonrenewable energy resources and most of its mineral resources.

Implicit Costs

Many of the interventions reducing resource demands or residuals outputs, by reduction or substitution, are effective not because of the materials they contain, but because of the resources used and of the residuals generated in their production. Each kilo-calorie of food energy that is produced in American agriculture represents the expenditure of five kilocalories of fossil fuel energy (Perelman, 1972), and the dissipation of significant amounts of biocides to the environment. Hannon (1972) estimates that 58,000 BTU are required to produce and distribute the 16 oz. throw-away bottles used for a gallon of soft drink. Returnable bottles would reduce this energy input to 19,200 BTU per gallon, while remelting and reprocessing the glass would require 62,600 BTU per gallon. The implicit energy, materials, and pollution costs associated with services as well as goods are the true measures of the effectiveness of interventions to conserve energy and materials and reduce residuals.

If the effectiveness of such interventions as product specification changes, reduced consumption, and product recovery is measured only in terms of the amount of energy and materials included in each product, then: 1) services (which involve delivery of information, not product) will not be considered; 2) the ramifications of water and energy demands, implicit in choice of materials, will be ignored; and 3) the material resources and residuals effects of electricity and water will be neglected.

Effects on Space and Density

Some interventions that affect energy and materials movement can have important effects on urban spaces and density.

Structures for treatment, storage and recycling or reclamation of residuals occupy space. A striking example of a treatment structure is the installation of 168 miles of 200 ft. wide cooling canals to condense exhaust steam from the Florida Power and Light Company's Turkey Point power plant (Anon., "Currents", 1973). If these structures create a nuisance, they will preempt additional space by restricting its use. Thus, the generation of power is the trade-off for space that is made unavailable or is of restricted value because of the effects of the residuals discharge.

Production changes involving energy and materials may increase or decrease space requirements. Decreased fossil fuel requirements per unit of agricultural product would probably necessitate additional land, farmed less intensively.

Other energy and material changes are also related to urban space or density. Opportunities to recover waste heat from electrical generation or solid waste combustion are better in dense urban settlements where the heat needs to be transported only a short distance.

Changes in transportation systems are more closely related to urban design. If the money invested in private automobiles were available to develop effective public transportation systems, urban density and design could be altered radically. The automobile has encouraged the dispersed nature of North American cities. Dense development today is commonly equated with reduced quality of life.

If automobile fuel were to be rationed, we might learn whether or not the association between increased energy input to urban systems and greater dispersion of urban areas can be applied conversely, i.e., whether reduced energy input results in denser urban development.

Interventions involving the increase of resource extraction or of environmental assimilative capacity require commitment of the natural environment for those purposes. A farm, which increases the capacity of the environment to yield food, a dam, which yields water or hydroelectric power, a sanitary landfill, or a river segment devoted to effluent disperson or in-stream aeration preempts use of that space for another purpose. The environmental change may not be entirely negative—pastoral scenery is aesthetically pleasing and artificial lakes may serve many purposes.

Space and density problems need further examination. Can current trends be reversed? Can more effective and satisfying cities be developed by planning high quality dense developments which locate people closer to their places of work? Can such development be achieved by utilizing the natural environment for recreation and agriculture and by providing transportation systems designed specifically to meet the needs for access to jobs at short distances, access to natural environments at medium distances, and the needs for long distance travel for business or pleasure (Royal Architectural Institute of Canada, 1973; Passonneau, 1972)?

Effects on the Individual

If we have accurately portrayed the way an individual makes decisions affecting energy and materials movements, then the interventions which appear most likely to be implemented are those with the least effect on the satisfaction the consumer derives from the quality or quantity of goods and services he receives. Some interventions do not directly influence the consumer; others directly affect his preferences and opinions.

All interventions have indirect consequences which may influence the decisions of an individual or a political-economic sector of society. Some interventions, such as the treatment and storage of residuals, can involve expenditures by industry or government that will decrease the consumer's buying power because of higher taxes or higher prices.

Changes in product specifications may go unnoticed by the consumer if the product's utility is not affected by the change. Changes which improve product efficiency, durability, ease of repair or cost of operation may be welcomed if those changes are not offset by unacceptable price increases. If the consumer reacts unfavorably to altered goods and services, that dissatisfaction may be expressed through the ballot box or in the market place.

To reduce consumption, individuals generally have to forego the satisfaction derived from goods or services or accept a substitute. Exceptions are cases where consumption can be reduced by using product specifications which provide better efficiency or longer product life without changing the utility of personal capital goods, or by substituting goods and services which give equivalent satisfaction but have less environmental impact.

When the consumer must choose between reduced satisfaction from a product and reduced satisfaction from the environment, circumstances favor the product. The satisfaction derived from products is more tangible and apparent at the moment of decision. It is continually reinforced by advertising. Information about environmental consequences, on the other hand, is often obtained by hearsay, may be inaccurate and is frequently late. Even the concerned individual must make his own estimates of the impact, if any, his decision will have on environmental quality. Implicit costs are difficult, if not impossible, for the individual to recognize and evaluate. For example, a consumer can consider the economic savings associated with conservation of electricity. The fact that two additional units of energy are conserved for each unit of energy he saves may be more difficult for him to appreciate.

The substitution of public services for private capital, especially in transportation systems, involves problems of providing equivalent service. Waiting in the rain for a slow crowded bus is not an adequate substitute for the convenience of the private automobile. The individual has no assurance, in forgoing his car, that equivalent service will be available, or that it will be reliable. Consumers are apparently willing to pay high prices, economically and environmentally, to retain the benefits of private automobile. Real alternatives are difficult to offer when urban form and density encourage private transportation.

ALTERNATIVE STRATEGIES

Motivation or incentives may be required to overcome the obstacles preventing application of possible interventions in the energy and materials flow system. Government can provide motivation in four situations: 1) When basic knowledge is lacking, incentives are needed to encourage research. 2) When available knowledge is not being applied, incentives are needed to encourage the development of technology. 3) When existing technology is not being applied, education is needed to disseminate the existing information. 4) When the available technology is known but is not being applied, incentives are needed to encourage its application.

Government can take such action: fiscally, by application of tax incentives, subsidies, or expenditures on research, demonstration projects or facilities; legislatively, by requiring or prohibiting actions that will implement decisions; or through persuasion, by education or advertising.

These actions can influence decision-making in the governmental, corporate or private sectors; all are interrelated. Private decisions are influenced by advertising, which may originate with government, industry or concerned groups of individuals.

Public and corporate decisions on interventions can involve waste treatment and storage, production changes, product specification changes or the reclamation of residuals. Individual decisions, on the other hand, are related to rates of use or consumption of products and to some kinds of reclamation such as recycling bottles and cans. Individual decisions do influence decisions about product specifications by affecting product consumption rates. Corporate decisions about a product's durability and efficiency, in turn, affect the consumer's buying habits.

We have identified types of action that government can take to create incentives for the implementation of interventions. Our review of interventions and of possible governmental actions is not exhaustive. The strategies outlined are not original ideas; all have been proposed by others. We have made no attempt to confine our discussion to problems which are distinctly urban. Some strategies we have advanced will affect energy and materials flows in urban systems, whatever the level of application (e.g., oil policy). Some strategies, applied at the urban level, will indirectly affect energy and materials movements beyond the borders of the urban system.

The discussion and recommendations that follow are frequently related to energy conservation. We are concerned about rapidly diminishing energy resources and rapidly increasing energy demands; we are concerned also that too many proposals or solutions are based on increasing supplies rather than on reducing demands. Makhijani and Litchenberg (1972) show that the standard of living (represented by per capital gross national product) is correlated with per capita energy consumption; but there are variations among the energy consumption levels of countries at similar GNP levels. Makhijani and Litchenberg estimate energy conservation measures could reduce U.S. per capital energy consumption by 37 per cent by the year 2000. If the U.S. could achieve an energy efficiency (energy consumed/GNP) corresponding to that of Sweden, France, Finland, and New Zealand, the GNP of the United States would remain at the present level.

Policies

If governments are to take actions capable of altering material and energy flows, they will have to commit themselves to the principle of resource conservation and residuals reduction and will have to adopt appropriate policies. Public decision-makers must realize that neither sources nor sinks are infinite.

When policy-makers are developing strategies to reduce resource depletion and residuals discharge, they should consider: the complete system of energy and material flow, including resource extraction, product generation and consumption, and residuals discharge; the socio-economic-political system; and the implicit costs associated with interventions. Policy-makers could encourage application of existing technology to reduce resource depletion or residuals discharge. Presently, such application is hampered by inertia, cost, and ignorance.

Education

Educating producers and consumers can help stimulate awareness of the nature and magnitude of particular problems and can indicate possible solutions. Educational programs designed to create such awareness should do more than indicate that resources are scarce and that residuals create unpleasant environmental effects. Attention should, rather, be directed to the amount and nature of increased consumption and its impact on particular problems.

Information about the relative efficiencies of products such as air conditioners or automobiles, or about the value of insulating the home, can help individuals make conservation-oriented decisions. Advertising appeals to the individual in terms of the satisfaction he will derive from a product. There is little information to offset advertising appeals by showing the adverse effects arising from the resource extraction or residuals discharge associated with a product. A disproportionately small effort is made to create demands for recycled or more energy-efficient products, or to change those attitudes resulting in the discarding of useful products simply because they are no longer new.

Design professionals can help supply information about resources and residuals problems and solutions. Both the American Institute of Architects and the Royal Architectural Institute of Canada have emphasized the need for architects to become more aware of energy conservation as an important design parameter. Both have encouraged the dissemination of information on energy conserving design techniques and on the implications of alternative design decisions in terms of energy consumption (Stein, 1971; Royal Architectural Institute of Canada, 1973).

Implementation

Besides encouraging and regulating the activities of other sectors of society, governments can significantly reduce energy and materials flow by improved product specifications for their own structures and equipment and by more efficient practices. Such policies can serve as valuable demonstrations of the government's serious intent. Actions taken under these policies show how interventions can be implemented.

The construction industry might be stimulated to apply energy-conserving techniques. Building codes can be used to encourage efficient use of building materials, to regulate the efficiency of air conditioning equipment, or to impose performance requirements (in kilowatts per cubic foot of space) for the energy consumption of buildings. Present incentives encouraging construction of buildings with high operating and low capital costs could be removed (Stein, 1971).

Significant savings might result from better building designs and specifications and from more efficient operation of building equipment. The insulation industry estimates that if all residential units were adequately insulated, 15.3×10^{15} BTU could be saved over the next ten years (Ball, 1973).

More efficient heating and cooling equipment within buildings also offers possibilities for energy conservation. One study of the efficiencies of 20 air conditioners with capacities of 10,000 BTU/hour revealed that their electricity demands varied from 800 to 2,000 watts (Hirst and Moyers, 1973). Another study comparing the costs of energy consumption in 24 Ontario schools indicated that the energy consumed per student per year increased as the age of the schools decreased. Costs for conventional schools with oil heat and large windows ranged from six to 20 dollars per student per year; the cost of an electrically-heated, air conditioned building was 30 dollars per student per year (Van den Hazel). A study of schools in Nova Scotia resulted in similar conclusions (Booth, 1971). Studies of a prototype 50-story office building, which would use 36.8×10^6 kwh each year if designed to conventional standards, revealed that changes in lighting, cooling, etc., could cut annual energy use in half (Stein, 1972).

Consumption of electrical energy could be reduced if price structures favoring high volume users were altered. If energy prices are not frozen at unrealistic levels, they can be expected to increase in response to increased fuel, construction and residuals control costs. The additional costs would have significant affect on demand (Chapman, Tyrrell and Mount, 1972; Lincoln, 1973). Increasing prices on industrial energy by taxes or regulation could save 3.8×10^{15} BTU per year by 1980 (Office of Emergency Preparedness, 1972). There is a precedent for this approach. Water consumption rates have been reduced by increasing prices during peak use periods and by charging higher unit prices for volume use (McPherson, 1971). As for the generation of electricity, if construction delays were alleviated to allow for earlier replacement of older, inefficient plants, and if loads were shifted to off-peak hours in order to smooth out demand, the industry could save 0.5×10^{15} BTU per year (Office of Emergency Preparedness, 1972).

Industry can be encouraged to increase its efficiency. U.S. Steel already uses natural gas given off by its coke ovens to fuel other operations. The planes of United Air Lines taxi to the terminal on two engines rather than four; the airline has also proposed reduced cruising speeds to reduce fuel consumption. The Aluminum Company of America is introducing a new process that may reduce energy needs by 30 per cent. Most of these actions are being taken to save money or to avoid rationing (Anon., *Newsweek*, 1973).

Design can greatly influence product composition, efficiency, rates of replacement, and reclaimability. Establishing minimum efficiency standards for furnaces, air conditioners and appliances and requiring the energy consumption of appliances to be stated on name plates, price tags and advertisements could save 0.6×10^{15} BTU per year. Incentives could also be provided to discourage unnecessary packaging, to reduce the input of scarce high energy or highly polluting materials and to encourage the manufacture of longer-lived products.

According to the Office of Emergency Preparedness (1972), increased recycling and reuse of materials could save 0.2×10^{15} BTU per year. The League of Women Voters has identified several obstacles to recycling in the United States: prices and taxes favor the use of virgin materials; shipping rates and service impede salvage operations; and laws and regulations hamper recycling. The League (1972) recommends several actions to encourage recycling: levy new taxes or remove existing tax benefits to shift price advantage away from virgin materials; subsidize the use of secondary materials; and build demand through changed government regulations.

"Direct recycling of large portions of industrial water is extensively practiced and is expected to grow in extent because of the combined effects of increasing source competition and upgrading of water pollution control regulations" (McPherson, 1972).

New freight handling systems, mass transit systems and improved urban design could save 6×10^{15} BTU per year (Office of Emergency Preparedness, 1972). Governments presently spend little money on the construction of railroads, the most efficient form of transportation, while they continue to make heavy expenditures for highways and air transportation (Hirst and Moyers, 1973). The energy efficiencies of transportation systems vary widely. Freight transport by truck, for example, requires 2,300 BTU per ton-mile. Railroad transport requires 680 BTU per ton-mile (Hirst, 1973). Reversing the current trend toward less efficient carriers (trucks instead of railroads, planes and automobiles instead of buses and mass transit) would yield significant savings.

Individuals can be encouraged to purchase products that require lower inputs of energy and materials, consume less resources, or generate less troublesome residuals. Taxes based on the size or efficiency of an automobile could encourage use of smaller more efficient vehicles. Automobiles may be labeled to indicate their energy efficiency. Vehicle design could be altered. The total energy efficiency of electric vehicles is about 20 per cent, only 10 per cent for internal combustion engines (Grimmer and Luszyczynski, 1972). The abandonment of private automobiles in favor of mass transit will depend on effective urban design, restrictions on automobile use, improved service in public systems and the effects of rising gasoline prices, shortages or rationing.

Research and Development

Governments can encourage research and development by financial support of work carried out by government, private agencies, and universities. Governments can also motivate corporate research by fiscal incentives. They can also explain how research and development may payoff in terms of demands for products and product changes which result in energy and materials conservation and reduced residuals discharge.

The purpose of research and development is the acquisition of knowledge about the operation of the system and its components. That knowledge can be used to develop better ways of manipulating the system and to evaluate alternative methods of altering the system. Manipulation of the system involves not only intervention to change the physical system, but also socio-economic modification necessary for the implementation of change. Evaluation of alternatives involves both consideration of an alternative's effectiveness in terms of direct changes in a resource or residual to which it is addressed and consideration of total effects: direct, indirect and implicit.

Evaluation of alternatives implies the existence of criteria against which performance of various interventions can be judged. Applicable criteria include effectiveness, cost to consumer, and ease or difficulty of implementation.

Research that may provide the knowledge necessary to develop better methods of manipulating the system of energy and materials flows, and that may evaluate relative performance of alternative interventions, can be identified in terms of the flow system we have outlined.

1) **Developing the Technology to Implement Interventions.** Many interventions will require additional technological development. Further attention needs to be given to: increased resource extraction capabilities, improved efficiency in the use of materials and energy and the generation of residuals in production processes, altered product specifications, reduced private consumption, and resource recovery and reuse.

Exploration for new energy sources and for more effective ways of extracting known resources from the environment should be continued. "Many unconventional sources are at a disadvantage in this respect, and the problems of bringing them into actual use may in some instances—for example, solar heating in homes—be much more difficult than developing the technology" (Hammond, 1972). Technology for tapping geothermal energy sources and for coal gassification and desulphurization techniques must be developed.

Improvements in the production of goods, services and energy should be encouraged, including waste product recovery, recycling, or process changes to reduce input of critical materials or substitution for them. The recycling of municipal water supplies may depend on the results of experimental programs such as that of the Denver Water Board, aimed at construction and operation of a full-scale water recycling plant by 1985 (McPherson, 1972).

New techniques and products to conserve energy through changes in building and equipment design should be encouraged; demonstration projects should be carried out to evaluate experimental assemblies and the effects of design, financing, tax and regulatory alternatives.

Products that are more energy-efficient, use fewer scarce resources, or have less serious residual effects should be favored. Work on products that reduce individual consumption through greater durability or more efficient use of electricity, fuel and water should be supported.

2) Research on Indirect and Implicit Effects. If rational decisions are to be made between alternative interventions in the materials and energy system, we must have enough information to evaluate their consequences in terms of the whole system.

"Before officials and citizens can objectively compare the cost of using resources in various ways, they need data on total costs, tangible and intangible, for producing, distributing, using and disposing of materials. Nor is there any way to trace the energy and water and manpower used or the liquid, solid and gaseous waste generated in producing, transporting and consuming any material from its source to its final disposal point" (League of Women Voters, 1972).

Altering consumption changes the whole pattern of energy and materials inputs to production and product delivery. For some products, effects may be relatively easy to describe. Reduced consumption of one unit of electricity, for example, represents a reduction in measurable quantities of raw materials and residuals. The effects of many other decisions are more complex and have not been adequately described. Several groups have made useful contributions in this area: Bower (1973) and his colleagues at Resources for the Future have looked at some of the inputs and residuals associated with the paper industry. Hannon (1972) has worked on the energy implications of bottle recycling. Makhijani and Litchtenberg (1972, 1973) have researched the amounts of energy used to manufacture a variety of products as well as residual energy use. Perelman (1972) has described inputs into agriculture. Healy and Dick (1972) examined the energy costs of the Bay Area Rapid Transit System. Hirst (1973) worked on transportation energy consumption.

More consideration should be given to the implications (in terms of energy, material and space requirements) of increasing, decreasing or stabilizing population growth rates. Contributions by Ayres and Kneese and by Odum (1970) are already available.

3) Research on Environmental Effects and Social and Economic Implications. Better predictive techniques are needed to assess the environmental impact of decisions involving energy and material movement. One research project which considers pollution, transportation, consumption patterns, housing, industry and agriculture is a regional energy use study being conducted by the Institute for Environmental Studies at the University of Wisconsin (Bryson, 1973).

Dorney (1973) and his colleagues have been working to "... develop a rapid and low cost method to inventory urban natural resources and urban amenities". Spofford, Russell and Kelly (1972) have been developing regional residuals management models "designed to deal with air and water quality and solid waste problems simultaneously because of the trade-offs among airborne, waterborne and solid residuals implied by the conservation of mass and energy". Their work recognizes the need for information "on methods of residuals handling and disposal, residuals reuse systems, production processes, alternative factor inputs to production, residuals generation associated with each production process, and finally, the externalities associated with production, reuse and disposal" (Spofford, 1971). Folk and Hannon at the University of Illinois have developed an input-output "energy, pollution and employment model" for 367 industries. Hamel and Brown (1972) have described the utilization analyses of energy systems and the development of an urban data base that uses Philadelphia as an urban laboratory. Regional environmental models are being developed at the Oak Ridge National Laboratory (Craven et al., 1973). An "Inter-institutional Policy Simulator" at the University of British Columbia is intended to involve the University with all levels of government in developing a model of the greater Vancouver region. Sub-models embrace economics, land use, human ecology, pollution, population and health systems (Vancouver Regional Simulation Study, 1971-72).

4) **Research on Public Perception and Communication of Environmental Change.** The nature and magnitude of imperfections in the systems for assessing environmental change and of time lags in the communication of information need to be examined. Koenig, Cooper and their colleagues at Michigan State University (1972) are studying the functioning and management of socio-economic control systems, including increased control over the energy requirements associated with system complexity.

5) **Research on the Value Judgments of Individuals.** The factors (prices, utility, advertising) which influence the relative values individuals place on products and the environment should be examined further. The effect of prices upon demand for water has already been considered (McPherson, 1971). Some work has been done on prices and demand for electricity (Chapman et al., 1972). More work is needed to describe the ways in which energy is used, similar to Makhijani and Litchtenberg's study of residential energy use (1973).

The relationship between consumption patterns and life styles also requires examination. Corr and MacLeod (1972) have described the effects of communal living on energy use—both in terms of use of capital goods and of altered life style. Life style is already affected by our current consumption patterns through power blackouts, gas shortages and pollution. We need, however, adequate descriptions of the relationships between various levels of consumption and corresponding alternative life styles. Commoner (1971) and his associates have made important contributions in this area (Commoner et al., 1971; Ehrlich, Heldren and Commoner, 1972).

6) **Research on Information Communication and Institutional Response.** We need to know how individual opinions are communicated to political decision-makers and how those opinions are weighed against other influences. Further, we should know how decisions are best implemented to bring about changes in energy and materials flows.

7) **Research on Incentives and Their Effectiveness.** In some cases, as with the recycling of beverage containers, the technology is developed, the energy advantages are apparent, and the price incentives are available, but incentives for energy and materials conservation are not being offered. We need to discover whether regulatory or market incentives are more appropriate tools for implementation. The problems associated with transportation systems are more complex and more far-reaching; methods to overcome jurisdictional and financing problems and consumer resistance need to be developed.

Geographical boundaries cannot define an urban materials and energy flow system. The system includes the movement of energy and materials from the point where resources are extracted to the point where residuals are discharged. The total system, driven by the urban population's demand for products, includes the materials and energy flow system, the natural environment affected by resource extraction and residuals discharge and the socio-economic-political system controlling energy and materials movement.

Resources are growing scarcer and residuals are creating environmental problems; accordingly, the resources coming into and the residuals flowing out of urban systems must be controlled. The disutility associated with resource extraction and residuals discharge can serve as a measure of system performance. We have identified seven categories within which disutility can be reduced. Five categories involve interventions in the energy and materials flow system.

When knowledge and technology are available, interventions to reduce energy and materials flow can be aided by governmental application of fiscal, legislative and persuasive incentives. Research can lead to better technology and to knowledge of methods for implementation. It can further provide better understanding of the consequences of alternative actions.

Effective methods to produce significant reductions in resource use and residuals discharge are already available and should be implemented, particularly to conserve energy. Many techniques which can be adopted without significant changes in life styles could improve the quality of urban life.

Research aimed at developing technology and gaining better understanding of environmental effects and of individual and institutional decision-making processes is necessary. The consequences of decisions may be more fully appreciated, as a result, and better methods of implementing policies for resource and residuals control may be developed.

INDICATORS OF URBAN ECOSYSTEM HEALTH

Social indicators may be defined as those measurable properties of a conceptualized system which contain socially significant information. They are explicitly or implicitly linked to social aspirations and can, to some degree, serve to measure progression toward or regression from socially desirable events, states or processes. Social indicators may provide information on more than one societal goal; however, a particular indicator seldom has an exact correspondence to a particular social aspiration. Likewise, more than one societal indicator may be used to assess a particular goal. Because indicators can—without identifying the social aspirations being considered—imply social significance where there is none, choice, interpretation and design of indicators may be critical.

We attempt here to define a more logically consistent model of social indicator systems than has been defined in the past. Our emphasis is two-fold: man-nature interactions within large urban populations, and the application of ecological principles to urban problems. This two-fold emphasis provides points of focus; it does not set the absolute limits of our discussion.

We first address fundamental issues of methodology, emphasizing the role of environmental-social indicators within the larger urban system. We then develop a flow chart to indicate the role, location, criteria and memory of indicators. Next, we design, analyze and evaluate criteria for, and preliminary means of, selecting indicators. We also discuss the aggregation and methodological problems associated with overlapping categories, including indices of resiliency, stability, diversity, variety, efficiency and controllability. Although it is too early to judge whether urban ecosystem research will become a discipline with a scientific basis, we presume that, given sufficient focus and a systematic taxonomy, it could. Our intent is the identification of a rather complete systematic taxonomy of social indicators.

For the purposes of the following discussion, we define an ecosystem as a complex and organized whole; it consists of a set of interactions between hierarchies of goal-seeking and adaptive open systems. These interactions occur within the system itself and between the system and the environment and set constraints on the system within the constraints of a general, inorganic environment. Each subsystem of the ecosystem may be all or part of every other subsystem.

An open system exchanges and transforms matter, energy and information with an environment. The boundary between system and environment is the result of decisions made within or outside the system. We are concerned here with those decisions made by an investigator outside the system.

The investigator decides whether to regard an ecosystem as: 1) a set of entities with properties; 2) a whole that is greater than the sum of its parts; or 3) a set of relations between relations. In the first instance, the whole is simply the sum of the properties of its entities. In the second, the whole is the sum of the properties of its components plus their relationships, including "secondary" qualities, "accidents", etc. In the third, the relationships constitute the components of the whole: form and organization are primary, content is secondary.

Information may be described generally as a pattern of variety carried by matter-energy markers. There is no intrinsic difference between one pattern of variety (information, in this instance) and any other (energy gradients, patterns of matter, noise). Information becomes information only when its pattern is recognized by a receiver. Information, therefore, requires a receiver, or further, two sender-receivers who are capable of perceiving patterns and differences and who can orient themselves in relation to their perceptions and to each other. In general, all ecosystems employ both continuous communication (feedback) and sporadic communication. Sporadic communication requires the receiver to select and integrate information from a discontinuous signal or from a continuous signal in which boundaries have been introduced.

Any intervention into an ecosystem will "punctuate" its signals in some way. It will impose an order which is not intrinsic in the system. Any decision by an observer to perceive an aspect of the system as information rather than as noise—for example, choosing an indicator and processing the information received from it—will also punctuate the system.

Choosing an indicator requires a goal, or a hierarchy of goals. Consequently, the choice of an indicator necessarily gives the indicator larger significance. If the indicator is described as a communication, the choice of it will involve decisions about how you wish to speak, about what you will speak, and about to whom you will speak. The question of "subjective bias" within a given discourse is almost irrelevant when compared to the significance of choosing between different discourses or ways of speaking, or between different levels of discourse.

The differences between kinds of discourse may make translation difficult or impossible;

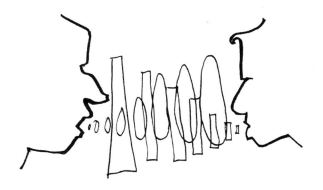

or those differences may be too great to be perceived, again resulting in faulty communication.

Differences between kinds of discourse become especially relevant when: the indicator chosen allows distinctions to be made between adaptive and counter-adaptive behavior or orientation; the behavior of the system is counteradaptive; and the way one is talking about the system and the problems of the system also manifests a counteradaptivity which significantly matches that of the system.

Goals

Goals can be regarded as desirable ends already achieved or to be achieved. Hypotheses, on the other hand, can be regarded as attempts to formulate goals. In socio-economic systems, hypotheses cannot, in general, be proven true or false; they can only become true or irrelevant. Indeed, any theory is "true" if it is sufficiently isolated from the context to which it refers and in which it arises. Given the complexity of the urban system, we may assume that, in some instances, the formulation and testing of hypotheses will lead only to trivial results.

Further, the history of human intervention into the natural and human ecosystems shows that, although an intervention may be made with the best of intentions, it often results in more damage than the damage caused by the situation it was designed to correct.

Long-range survival in any ecosystem is determined in large part by the system's uncommitted potential for change. Stable natural ecosystems have such potential. Consequently, in the selection of indicators, highest priority should be given to providing reliable ways to detect irreversible losses of flexibility in the system under observation.

Definitions of Indicators

The rationale for selecting indicators should, presumably, reflect society's goals. Choosing realistic expectations for a system as complex as ours requires pragmatic as well as theoretical options. We do not have, as yet, the information and the degree of citizen participation necessary for defining specific and operational goals for the urban system. Hence, instead of attempting to define specific goals, it may be more profitable to define what is undesirable.

We can identify regions which constitute ecological, economic, social and political irreversibilities. These are regions we do not want to enter. Our system's region of flexibility consists of that set of alternatives which are mutually viable in social, economic, political and ecological terms.

Structural and functional controls determine the complexity of the region of flexibility. Structural controls are those constraints which are embedded in the configuration of the system and which, by definition,

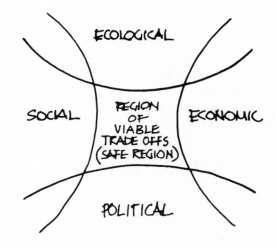

Figure 3 Zones of irreversibility

cannot be located or altered. Functional controls are those control mechanisms which can be modified or restructured. For example, the legal structure in the urban system, although it is integrated into the system's set of complex relationships, is logically exterior to the system's structural controls. Functional controls can be located and can be altered.

Within the boundaries of unacceptable irreversibility, the system follows a biased random walk strategy. The acceptable states are not purely functions of the boundaries but are related directly or inversely to the level and kind of organization of the system. The system's structural and functional control mechanisms set the boundaries of irreversibility. An anti-surplus economy, for example, will have a boundary quite different from a surplus economy; both boundaries will shift with the increasing input of matter and energy from outside the system.

Stability and resiliency may be defined in terms of a single static moment and in terms of a temporal continuum. The single-moment, static view puts systemic process and structural considerations in a timeless perspective. The continuum view allows examination of stability and resiliency over an extended period of time. For our purposes, we assume that increasing the stability of a system decreases the system's future potential for change, i.e. its resiliency. The converse holds true as well.

The relative freedom of the system to change is determined by the ratio of stability to resiliency: the greater the system's stability, the less free it is to change and to react to change; the greater the resiliency, the greater and more multi-dimensional will be the system's potential for change and accommodation. We call this stability-resiliency ratio the "morphostatic" freedom of the system—that freedom which is defined within the structural controls of the system by its configuration, and beyond which the system cannot go without changing its structure.

"Morphogenic" freedom, the second order of systemic freedom, is the ratio of stability to resiliency as it is maintained through time by the system's changing structure. The system can restore lost flexibility at one level by evolving to a higher level.

Irreversible losses of flexibility may be represented by the monoplanar random walk, where any trajectory moving toward the boundary zone raises the possibility of an excursion into the zones of irreversibility. When boundaries are being approached, neither the structural nor the functional control mechanisms of the system are adequate to return the system to a position of adequate resiliency or flexibility. Accordingly, the extrastructural or functional controls may have to be changed in a significant way—for example, by introducing a new order of control into the system.

Increased control at a higher level in the system may be exerted to reduce the resiliency of the system. A higher order of controls may increase the velocity of the trajectory before returning the system to a safe zone. The corrective action itself, however, may move the system into the zone of irreversibility. Repeated excursions toward the zone of irreversibility will make increased control easier to exert: the system will not only adapt, but will adapt to the process of adaptation.

If we assume that the extrastructural or functional controls are equivalent to superstructural political and legal controls, every step up the control hierarchy may, in itself, be irreversible as far as the morphostatic state of the system is concerned. Excursions into one or more zones where flexibility would be irreversibly lost would presumably generate social chaos. Continued movement up the control hierarchy could constitute movement toward totalitarianism.

Urban System—Natural System Parables

Our capacity to model natural systems and our consequent understanding of those systems lead many people to believe we can readily and usefully apply our experience with natural system to urban systems. Indeed, many people believe that techniques used to model natural systems can be used to understand urban systems and to plan their development and continued operation. The relatively untrammeled natural environment appears to exhibit a significant degree of orderliness and stability; many people assume that the processes embedded in natural systems may also operate in urban systems. It would seem that discovery and application of natural system "laws" to the urban system would result in order and stability in the city.

However, natural environments are not always orderly and neat from a human standpoint. The operational requirements of natural systems may be unacceptable to man. Are we, for example, willing to kill and eat our young to survive a period of food shortage as some animals do?

The principles common to both natural systems and human social systems are too poorly defined for us to determine when we should parallel natural systems to best serve human goals and when we should not.

It is misleading, and even dangerous, to attempt to superimpose all concepts derived from models of natural systems onto urban ecosystems. Rather, we can more pragmatically and fruitfully adopt the approach of the ecosystems scientist and view the city as an urban ecosystem with its own set of principles and requirements. A list of observations, concepts, principles and models of natural systems may well serve as a checklist for the study of urban systems. Apart from questions of the desirability or feasibility of applying natural system principles to the urban system, there is an additional question of whether it is indeed necessary to utilize those principles to ensure that urban systems survive and ultimately become "suitable" places to live. It may be equally useful to uncover the principles underlying the urban ecosystem itself, and to understand how structural and functional control mechanisms affect the functioning of the urban system.

Comparing Natural and Urban Ecosystems

Ecological theory attempts to understand and to represent the discernible patterns of necessary interactions between biological and nonbiological elements of an ecosystem. In natural systems, the strategy for survival aims at preserving whole elements, species or populations; individual elements are partly expendable. Currently, the strategy for urban systems favors the accumulation of wealth and high material and energy flows. Individual life is rarely expendable.

Selection Processes. In the Darwinian sense, fitness is an empirical test of the suitability of each individual's characteristics in relation to its environment. Selection proceeds within the structural relationships of the ecosystem. Natural selection is a feedback process that continually evaluates the changing environment of a population.

In urban ecosystems, economic, political and social selection processes are the dominant influences determining the system's structure and behavior. These process are not restricted to individuals or subpopulations and are not necessarily empirical. Nor are there always feedback control mechanisms.

Time Domain. The selection processes in urban and natural systems determine their respective time domains within which adjustments are made in the system. In natural systems, time domains are based on the generation time of organisms, governed by genetic selection and phenotypic responses to seasonal phenomena. Time domains in planning urban systems often span decades—economic horizons are viewed five to 15 years ahead and political time domains depend on terms of office.

Controls. Natural systems use continuous information or information from analogs to synchronize the behavior of individuals with each other and the behavior of individuals and systems with the physical environment. The multispecies control processes—competition, predation, territoriality—and the resultant evolutionary patterns of co-adaptation, on the other hand, utilize digital information, received and used discretely.

The urban ecosystem responds chiefly to digital information. No major urban system responds to continuous information, except perhaps the market system which responds to prices. The controls of natural and urban systems have little similarity.

Spatial Designs. Natural systems have populations whose interactions—competition, predation, etc.—set limits on the size of an area within which tight integration of organisms can be maintained. Natural systems, therefore, are highly distributive spatially; they appear on the landscape as a mosaic of biological subsystems.

Urban ecosystems respond to a different set of interactions—economic, political and social. Such factors as scale economies and urban cultural richness have increased the degree of "clumping" of organisms in the urban system. The system is highly concentrated spatially.

Materials Dynamics. The integrated flow rates and patterns of flows around which a natural system is organized accumulate sufficient available nutrients to maintain the complex biological community. Material flow patterns are predominantly internalized and flow rates are modest.

An urban system, responding primarily to flow rates, is characterized by its flow patterns. It is highly dependent on outside sources for its material resources and for the assimilation of wastes.

Values. Values, morals and ethics may not have a counterpart in the non-human world. Enhancement of life may be a human concept only, and may have no meaning in nature. In natural systems, weak members generally are not supported by other members of the population. In many human systems, though individuals may be unable to fend for themselves, they are supported for social reasons. Such behavior may contribute to the enrichment of human life, but it may also decrease the probability of survival of the overall system.

Fluxes. Comparisons might be drawn between transportation, storage and transformation of matter, energy, and information in natural and human systems. Many comparisons will reveal no inherent differences between the two systems. Each has living and nonliving components in each of its functional or substance categories. Direct application of principles underlying these similarities can be useful.

We believe major gaps exist in conceptualization and in identification of a methodology useful in designing a systematic set of urban ecological indicators. Problem areas include those below:

1. amplification through feedback processes;
2. the transmission of events in one part

of the system into unexpected events in its other parts;
3. the recognition of organic and social complexity and the inadequacy of present methods for detailed explanation;
4. the necessity for properly analysing values and behaviors of any given system;
5. the recognition that open systems are goalseeking and involve communication and control;
6. the distinction between energy-entity explanations and communication-relation explanations;
7. the recognition that the distinction between organism or system and environment is not a boundary that is equivalent to a barrier, but a point of communication and exchange;
8. dealing with relations between relations (i.e., between systems, environments and ecosystems);
9. the distinction between long-range and short-range effects;
10. the protective and stabilizing nature of redundancy in systems and the instability of efficiency;
11. the understanding that survival in nature involves not simply the fittest individual, but more significantly, the fittest ecosystem;
12. the distinction between causality and constraint, between "determinism" and relative structural freedom; and
13. the recognition that levels of communication and control can be distinguished logically.

INDICES OF CRITICAL PROCESSES AND STATES

The physical, biological and cultural links between the human component of the urban ecosystem and the natural system surrounding it are essential to the basic metabolism of both systems and, consequently, must be explicitly reflected in indices which are developed. Currently, only the links involved in the dynamics of material and energy relationships have been adequately examined.

The relationships between the urban and natural systems can be divided into subclasses representing the essential functions of transformation, transportation and control. Transformation refers to the sequential chain of production steps concentrating and structuring materials or storing potential energy. Beyond transformation lie the processes which result in the dispersion or disassociation of materials and the release of energy during the course of consumption. Transportation is a process of redistributing and storing the materials which affect the structure of the urban system in time and space. Control processes, if rationally designed, obtain, process and transmit information; they require relatively small amounts of material and energy resources.

An operational definition of an urban ecosystem includes in its specific set of components: resource extraction, materials production, human consumption, residuals handling and the metabolic transformation of residuals (see Figs. 4–8). The indices needed to follow the dynamics of the urban system are functions or correlates of: flow rates; directions of flows; opportunity costs in terms of energy and material resources; stability characteristics as behavioral properties; and risk functions associated with irreversible actions. Such criteria can be used to identify a necessary set of ecosystem indicators for physical and biological processes, but not to identify a set sufficient for the total system, since the social, political and economic dimensions of the urban system have not been explicitly included.

Processes: Interventions, Indicators, and Models   145

The definition of a sufficient set of parameters for choosing indicators will be a function of the system of interest and/or the constraints imposed by the environment. A

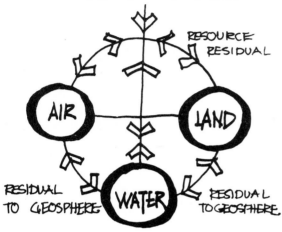

FLOWS: CONSUMER DEMAND

COSTS: ENERGY
NATURAL RESOURCE ALLOCATION
(RESOURCE OPPORTUNITY COSTS)

STABILITY: LONG TERM AVAILABILITY OF ADEQUATE CONCENTRATIONS

RISK: TOXICITY OF RESIDUALS

Figure 4   Resource extraction

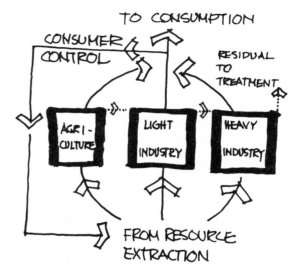

FLOWS: CONSUMER DEMAND

COSTS: ENERGY
MATERIAL FLOWS
LAND AREA TO PRODUCTION UNIT
H_2O RESOURCES TO PRODUCTION UNIT

STABILITY: FUNCTION OF RATES OF UNIT FLOW

RISK: STABILITY OR DEPENDABILITY OF OUTPUT
TOXICITY OF RESIDUALS

Figure 5   Material production

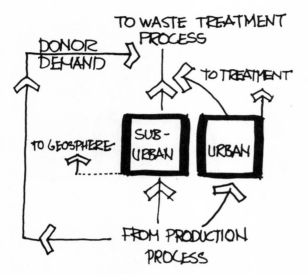

FLOWS: DONOR DEMAND

COSTS: ENERGY
PRODUCT FLOWS
LAND AREA OF HUMAN HABITAT
H_2O RESOURCES FOR HUMAN CONSUMPTION

STABILITY: DURABILITY OF GOODS
STABILITY OF PRODUCT FLOWS

RISK: TOXICITY OF RESIDUALS

Figure 6 Human consumption

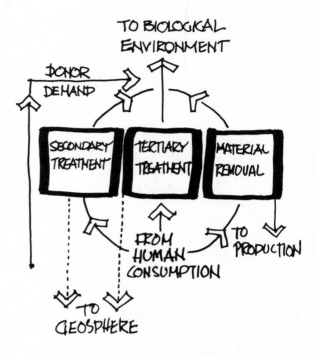

FLOWS: DONOR DEMAND

COSTS: ENERGY

STABILITY: FLOW RATES OF UNIT MATERIALS

RISKS: TOXICITY OF RESIDUALS

Figure 7 Residual waste treatment

logical framework, therefore, will be one identifying a sufficient set of indicators. Of the urban system's five components—resource extraction, material production, human consumption, residual waste treatment, and environmental biological assimilation, the natural system's capacity for assimilation of wastes is the most critical set of environmental constraints.

Decisions on allowable tradeoffs between the environment's assimilative capacity and

Processes: Interventions, Indicators, and Models   147

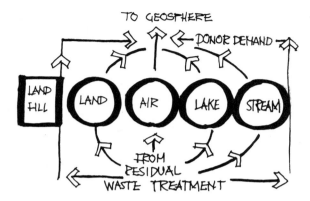

Figure 8   Environmental assimilation

forecasts possible excursions into zones of irreversibility, a richer set of information, based on mechanistic relationships, will be needed. These indicators for "crisis" conditions are more technical, less translatable, and more precise. Consequently, they are more costly.

Ecological Irreversibility

We have too little understanding of ecological systems to ascertain accurately whether or not wise decisions today will remain wise decisions in the future. Planners, therefore, need to minimize the decisions which may result in irreversible environmental changes. Several classes of changes may be functionally irreversible in the human time span.

1. Extinction of a species is an irreversible change. Eradication of a species is a planned effort to force its extinction. Eradication can have negative results, however; if we discover we need the genetic information of an eradicated species, we will be unable to retrieve it. Eradication of any species, therefore, is a dangerous practice and, even in the control of "pest" organisms, it must be discouraged as an environmental management policy.

2. Introduction of plants and animals from one region into another may have irreversible effects. The results of the introduction of the sea lamprey and alewife into the Great Lakes, the Japanese beetle to the East Coast, or quackgrass and dandelion throughout North America illustrate problems associated with such introductions. Similar introductions have often been made in urban systems.

3. Local or global contamination of the physical and biological environment with nondegradable chemicals constitutes an irreversible alteration. We lack the technology to remove contaminants from the environment without introducing additional ones.

4. Physical degradation of critical areas by

desired environmental quality define another set of constraints. These constraints in turn define the range of conditions under which natural homeostatic mechanisms can provide a sufficient set of systemic controls. While the system remains flexible, few external controls are necessary and indices with low information levels suffice.

However, when the trajectory of the system

such actions as the dredging of waterways, the damming of stream channels and the filling of marshes and estuaries can have irreversible consequences. The urban ecosystem planner should become sensitive to the environmental costs and the ecological risks associated with such physical degradation.

5. Depletion of concentrated resources by wide distribution of waste products is an irreversible loss. The reclamation and reconcentration of waste materials from the Gulf of Mexico, Lake Erie or even from a small sanitary land fill would require enormous amounts of energy. Phosphorus, a critical chemical, is already widely dispersed and reclamation may no longer be possible.

6. Committed use of critical environmental resources for purposes which involve large amounts of capital can cause essentially irreversible socio-economic changes. Placing a shopping center on agricultural soil or an airport above valuable deposits of sand and gravel may make such resources unavailable for other future uses.

7. Political and economic irreversibilities are linked to the structural configuration of the system. The most critical relationship is the co-evolution of the physical components of the system and their control mechanisms. Structural configurations evolve along a trajectory of increasingly larger time and space domains in response to economic selection pressures and technological innovations. As the size of each subsystems expands, the control functions or political design necessary for adequate regulation must shift from predominantly feedback or response controls to forward-looking or anticipatory controls.

Recognizing Brinkmanship

Given any group of system behaviors such as production, consumption or waste processing, and given a systematic way of defining environmental, social and technological constraints (see Figs. 9–13), it may be possible to determine the boundaries defining a system's "safe" region. The tradeoffs a society is willing to make to satisfy its own immediate demands on a system determine how far that society places itself from the boundaries of undesirable regions. When this "brinkmanship" policy is followed, however, the gains made in terms of high immediate utility are offset by the increasing costs of preventing excursions into undesirable regions.

The process of waste assimilation provides an example. Given a river system in a temperate region and a group of organic residuals, at a fixed temperature the river's maximum assimilation of material might be 100 lbs. of residual per mile per day. If society requires the river to process this total amount, however, the river's biological community will be altered; "undesirable" species such as trash fish, sludge worms or algae will thrive. On the other hand, if society wishes to maintain an abundant trout population in the river, fewer wastes should be introduced into the river's ecosystem for assimilation. Similar tradeoffs are involved in other urban processes.

Interactions Between Indicator Data Requirements, System Trajectories and System Controls

As a system changes, it may move into "zones of trouble". The system's state, its existing condition and structure, may be called a "state vector". The system's trajectory, its direction and rate of motion, may be called a "velocity vector". Possible trouble zones may be categorized as the environmental, the social, the economic or the political (see Fig. 3). Although no sharp boundary separates the safe region from trouble zones, there are absolute limits beyond which recovery is no longer possible, e.g. a global nuclear war.

Once the system moves into a trouble zone, dysfunctions begin occurring and system maintenance costs rise sharply. The additional costs for achieving system recovery may be termed recovery costs.

The more rapidly the system is moving at the moment the system enters a trouble zone, the deeper will be the excursion into that zone. Higher velocity excursion into trouble areas has several ramifications: the recovery costs will be greater; the number of control options for achieving recovery will be reduced; and the probability of recovery will be reduced, perhaps to zero, if human or other forms of life should disappear. These effects, however, are not linear and are probably not everywhere continuous; they are, therefore, difficult to predict.

Indicator data requirements, considered as a function of the system velocity vector as it moves toward a trouble zone and specified in terms of type, accuracy, precision and timeliness, are a function of: 1) the normal component of the system velocity vector (war, for example, will cause a deeper excursion than insulting an ambassador's husband or wife); 2) uncertainty as to the location and gradients of the trouble zones (in many cases, the effects of thermal pollution on the limits of waste assimilation capacity are still unknown); 3) the cost and rate of system recovery; and 4) the effects of the recovery controls which are to be applied (it takes much longer to recover from a change of government designed to correct a threat to security than from the appointment of a study commission).

There is interplay among specialization, biological development and economic production. To a biologist, development means the rejection of sets of viable alternatives or the constraint of options. To an economist, production efficiency means some comparative advantage. Both biological development and production efficiency require specialization.

In both natural and social systems, specialization optimizes production. The specialist benefits from a static and predictable environment. The generalist is more diverse and less productive. Generalization rather than specialization optimizes the ability to survive and enriches life; the generalist, thus, survives in a variable environment, especially one that is unpredictable. The demand for production may often promote specialization, yielding short-run advantages. The world, however, is both variable and hard to predict; specialization, thus, may lead to disaster. The generalist, on the other hand, maximizes both survivability and enrichment, two aspects of our stated goals.

Accordingly, some of man's strategies—such as single crop agriculture, urban migration, or a city's economic specialization—make the system more productive but less flexible and more disaster-prone.

A certain amount of self-generated noise will help the system survive. For example, if the genetic code of a species is replicated exactly, if there is no noise in the channel, mutation does not occur and the species will not evolve. On the other hand, if there is too much noise in the system, not enough of the genetic code will be transferred to the next generation and the species may die out. The more variable the system's environment, the more internal noise is essential. In this context, noise is system's insurance for survival; the cost of noise is inefficiency.

When a system wanders into a trouble zone: 1) maintenance costs begin to rise sharply; 2) recovery costs rise, perhaps discontinuously; 3) control options are lost; and 4) survivability may become threatened. Remaining control options may produce secondary excursions more severe than the primary one. Production efficiency must then be increased, but could be disastrous if system production is routinely maintained at an inflated level while operating in the safe region.

The system can be operated indefinitely in trouble zones as long as the dysfunctions are acceptable and the higher maintenance costs can be met. True costs, however do not always appear immediately and cannot always be met. The asbestos accumulation in a construction worker's lungs, for example, may kill him 30 years later; deep-well injection of wastes can eventually precipitate earthquakes.

A system is sometimes operated at unnecessarily high production levels which have high resource flow rates and inadequate reserves while still in a safe region. Why is this necessary? In our present situation, high production is demanded by the scarcity resulting from population pressure and by the general philosophy that "progress equals growth equals good".

When indicators are designed, data requirements need to be considered in the context of a dynamic system which has both state and trajectory vectors and of a system operating in a safe region surrounded by various trouble zones (see Fig. 3).

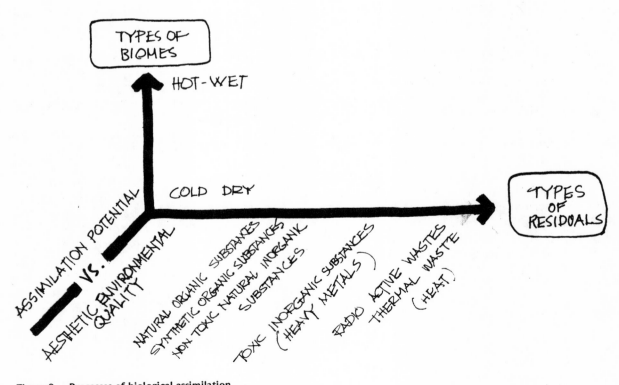

Figure 9 Processes of biological assimilation

Processes: Interventions, Indicators, and Models 151

DIAGRAMMING URBAN PROCESSES

The processes which constrain the urban system—biological assimilation, residual waste treatment, consumption, material production, and resource extraction—can be portrayed diagrammatically.

Figure 9 depicts the processes of biological assimilation. The vertical axis indicates the constraint, i.e. the assimilative capacity of various types of biomes. Hot, wet biomes have a greater waste assimilation capacity than do cold, dry ones. The horizontal axis shows the classes of demands made on the assimilative capacity. The third axis represents the trade-offs involved in controlling the process of waste assimilation; it poses the question how much environmental quality is to be sacrificed in the interest of waste assimilation. If the wastes we discharge are difficult for the environment to assimilate, we lose more aesthetic environmental quality than if the wastes we discharge are easily assimilated.

Figure 10 outlines the process of residual waste treatment. The vertical axis again indicates the constraint factors, the various technologically-feasible methods of residuals treatment, arranged in hierarchical order according to cost. The horizontal axis depicts the demands which can be made on residual treat-

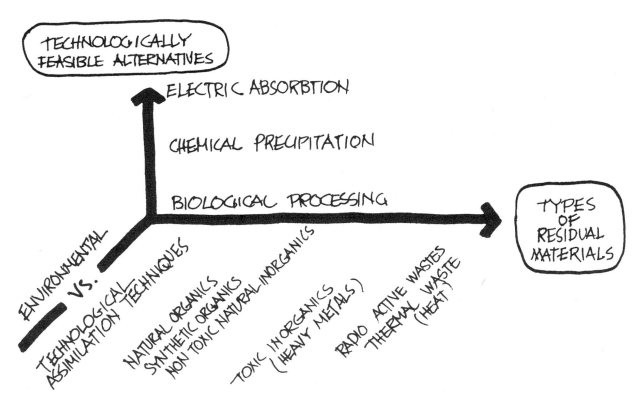

Figure 10 Residual waste treatment

ment technologies. If we are unwilling to pay the economic costs associated with waste treatment technologies, we incur other costs in using the environment as a waste receptacle. The third axis includes the combined use of residual treatment technologies and environmental capacity for dealing with waste. Knowing where we are on this axis makes the costs of using the environment as a waste receptacle more explicit.

Figure 11 diagrams the human consumption processes. The constraint is the set of mechanisms provided for allocation. The consumptive demands are arranged on the horizontal axis; they consist of all the items people consume. Different demands require different methods of allocation. For example, private firms cannot effectively control water pollution or provide clean air. The nature of the items we demand determines, to a large extent, how they will be distributed. Distribution, in turn, determines whether income will be evenly distributed among the members of the population or whether large amounts will be concentrated in the hands of a few. If allocations are made publicly, it is likely that goods and services will be more evenly distributed than if allocations were made privately. In examining human consumption processes, we need to know what the consumption-

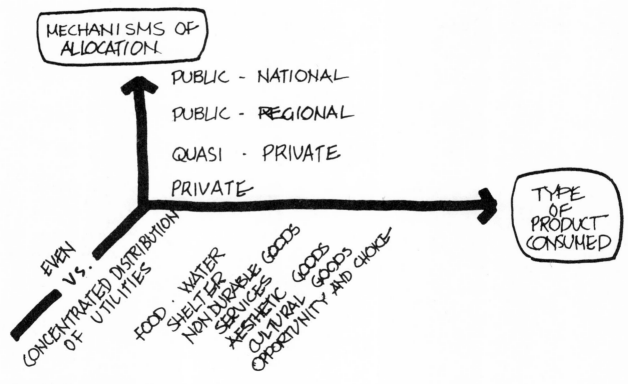

Figure 11 Human consumption

Processes: Interventions, Indicators, and Models 153

distribution processes imply for the distribution of well-being.

Figure 12 shows the process of material production. The pattern of demands is reflected in a mixture of productive techniques. Some techniques are human intensive; others are not and presumably require sizable amounts of capital. When we use non-human-intensive techniques, we more often take advantage of the efficiencies associated with large-scale operations. The result is a highly aggregative production process accompanied by substantial social costs not considered in consumption or production decisions. The third axis indicates the external environmental and social costs resulting from the interaction of consumption decisions with production constraints.

Figure 13 depicts resource extraction processes. Extraction is constrained by technology, the vertical axis. The question asked is, how much future flexibility we are willing tradeoff for immediate fulfillment. For example, if we drive resource extraction processes to the limits of the technological constraints, we will probably be giving up most of our future flexibility. If extraction proceeds at low rates, future flexibility is preserved.

THE OPERATIONAL FRAMEWORK OF SOCIAL-ENVIRONMENTAL ACCOUNTING

In structuring an operational system of social-environmental accounting and thereby understanding indicator systems in their most fundamental form, three problems must be

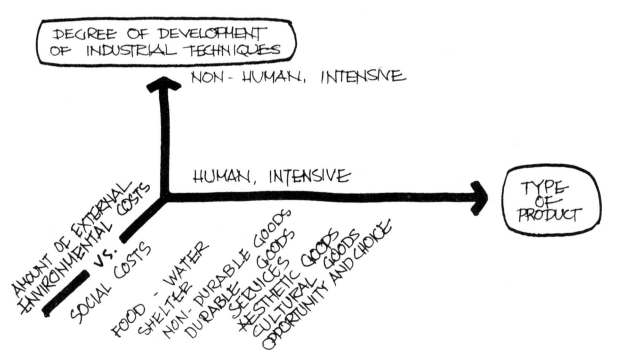

Figure 12 Material production

154 Part II

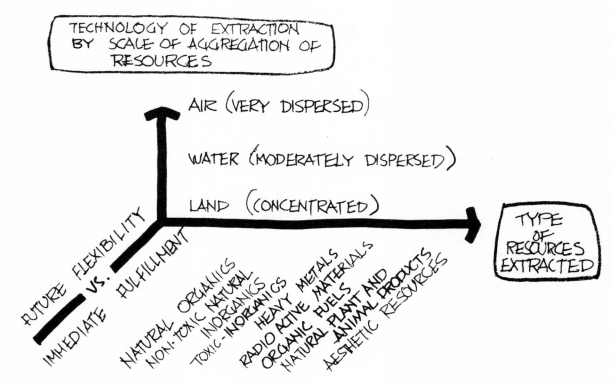

Figure 13 Resource extraction

addressed. First, principles and procedures must be developed for obtaining the data on which the accounting is based. Second, principles and procedures must be developed for processing the data. Third, the data must be disseminated and made accessible to those who would use it. We address each of these problems below.

Data Acquisition

A social-environmental accounting system is a set of elementary statements that are asserted to be true. If a system is to be operational, we must be able to test each statement by observation and/or experiment. For example, we can test the statement "The hourly average concentration of suspended particulate matter measured by the xyz method at the corner of 13th and Vermont Streets in Washington, D.C., on February 14, 1973, is 87 micrograms per cubic meter". The statement is obviously based on a relatively specific measurement and decision process.

However, adherence to such a concept of operationalism can be confining. Many objects of direct human concern, and often of overriding importance, are not easily assessed by experimentation or observation. The statements "People are alienated in Washington, D.C." or "The Washington, D.C., ecosystem is more resilient this year than last" certainly discuss phenomena which as directly and vitally concern residents of Washington, D.C., as do statements about the level of suspended particulates. Yet we have little knowledge about how to test such propositions.

If we fail to attempt to make important propositions operational, social-environmental accounts will be rendered sterile. We may try to surmount the difficulties of testing such propositions by defining related or associated concepts that lend themselves to observation or experimentation. For example, to assess the validity of the statement about alienation, we might make observations about violent crimes, complaints, demonstrations, feelings about the environment, etc. Clearly, there will be disagreement about what is related and what is unrelated to alienation. It would be prudent, in observation, to accommodate as many viewpoints as resources permit.

Concepts may appear difficult or easy to test; in either case, precise definition of concepts and specification of methods of observation or experimentation are essential for data acquisition. Precision is necessary to approximate the ideal of "statistical control".

The need for precise definition is imperative for all systems of accounting, but other problems inherent in ecosystems also deserve consideration in data acquisition. First, because ecosystems are constantly in flux, the conditions under which observations and experiments are made at a particular moment are likely to differ from the conditions of another time. Each opportunity for experimentation is unique and experiments sometimes fail; it may be necessary, therefore, to do other observations or experiments simultaneously. Otherwise, valuable data may be lost.

Second, since urban ecosystem processes are not well defined temporally, a continuum of observations and experiments adequately accounting for time and space variables may be desirable. However, providing for multiple and continual experimentation is costly and compromises based on probabilities may be necessary.

Third, some portions of ecosystems may be extremely sensitive to intrusion by observers. Processes are sometimes altered by the process of data acquisition. Observers should take care to prevent such intrusions when possible, or at least to understand the effects of intrusions.

Finally, any accounting system of acquisition must be capable of adapting quickly to change since attitudes toward and goals for urban ecosystems are highly fluid.

Data Processing

We consider, here, only two general principles. First, control and consistency checks are desirable in data processing. Often, erroneous data will be reported or correct data will be misreported in the processing system. Multiple observations, when possible, and well-defined relationships among different observations of phenomena can be used to check the plausibility of individual observations. Errors should be minimized.

Second, data processing operations themselves are subject to error. Data processors should take steps to reduce such error and to insure that any remaining error is compensated for through statistical controls.

Data Dissemination

To be useful, data must be available and compilable. Data banks, accordingly, must be structured. Raw observations can be considered the leaves of the "data tree"; aggregations of data constitute the twigs or larger branches. If the prospective user is to move from evaluations and conclusions about data to raw observations themselves, the data must be expressed in language that is understandable. If different interests are to be accommodated, the access system must be left open. With such an open system, tampering should be prevented by ensuring that the user can only revise evaluations of the data and not the data themselves.

CRITERIA FOR SELECTING INDICATORS

The selection of indicators to show the status and direction of change in the urban system requires an approach that is both idealistic and pragmatic. We need to describe the ideal properties indicators should have, but we must recognize pragmatically the difficulty inherent in collecting and disseminating information about a complex system.

Indicators are meant to describe the state of the system, either in terms of its current condition or of conditions likely to occur in the future. Present and future conditions may arise from actions taken by entities within the urban ecosystem or from events beyond the control of those entities.

An ideal set of social indicators might have these elements: exhaustiveness, independence, perceptibility, relevance and rationality.

Any set of indicators, however, cannot have all elements equally; if the set is to emphasize one element, it must reduce any other element with opposing characteristics. For example, the elements exhaustiveness and independence may conflict. An exhaustive set of indicators allows for the description of any condition or information which could be used as a social measure. Not everyone can be involved in creating an exhaustive set, however, and different people recognize different descriptions of a given phenomenon. Thus, a set of indicators cannot be truly exhaustive, though it may appear so to those creating it. Further, exhaustiveness may create a bias if one piece of information is represented by several indices and another piece by only one index. For that piece represented by several indices, partial overlap occurs and the indicators cannot be considered independent. For example, the indices of biological oxygen demand and dissolved oxygen both indicate the biological state of a body of water. There are models which allow one to determine what the dissolved oxygen will be if the biological oxygen demand and other factors are known. One might choose to report these factors, but the user has to know how to process the information if he is to determine dissolved oxygen. Although there is reason to use overlapping indices, there is reason to have independent indices as well.

A partial solution to the conflicting criteria of exhaustiveness and independence might be the construction and implementation of an appropriate information hierarchy, including the use of agglomeration weights which can be varied according to user.

The concepts of perceptibility and relevance may also conflict. Perceptibility can be defined as that information which can be recognized and evaluated by a layman. Relevance, on the other hand, means that an indicator is reasonably accurate and precise in describing a phenomenon; accuracy and precision must be determined by professionals. These elements do not always conflict; both professionals and nonprofessionals understand measures such as temperature and precipitation. With other measures, however, particularly those relating to system ecology, professional expertise may be necessary for understanding.

A further criterion for an ideal set of indices is that measurement be based on derived rather than associative measures when possible. Derivation involves the specification of a logical structure; association may not. Associative measures would be used, however, if they were the only ones which could be estimated.

Urban ecosystem planning is ordinarily done with so little information that the dichotomies between indicators we have discussed are usually of little concern. Few serious attempts are made to project future states of the urban system. For example, the indices estimated when the decision was made to install sanitary sewer systems in Suffolk County, New York, were: 1) the cost; and 2) the probable fall in the ground water table. Indices left unexamined included: 1) the current rate of dispersion of nitrates in the ground water aquifiers; 2) the net effect on the county's economy; 3) the changes in vegetation due to the falling water table; 4) the effect of secondary treated sewage on shellfish in the Great South Bay; and 5) the change in the amount and kinds of recreational land available in the county. For the most part, these indices are neglected for either of two reasons: 1) inadequate knowledge about how estimates should be made; or 2) the cost of estimation was higher than the planning funds available would allow. Sufficient knowledge and sufficient funds for developing a reasonably inclusive list of indicators appear to be two pragmatic criteria in considering the use of indicators.

All the criteria for indicators discussed above are affected by scale. In general, those fitting on a ratio scale are the most desirable (and the most expensive). Other scales which are less expensive substitutes include: 1) rank scales; 2) dominance scales (rank scales for which transitivity cannot be guaranteed); and 3) a scale reflecting the presence or absence of a condition.

The final criterion of the acceptability of an indicator is: the information must be collectable within the mores of society and must be presentable within the boundaries of societal understanding and interpretation. Acceptability and presentability may not create difficulties if one is dealing with other professionals, but the acceptability or presentability of a particular indicator may cause a layman to reject an entire set of potential indicators.

PROBLEMS OF AGGREGATION AND DISAGGREGATION

Aggregation and disaggregation cause problems in any attempt to measure and describe natural and social phenomena. A principal difficulty in developing a set of measurements which can be aggregated or disaggregated stems from the fact that disaggregates, when summed, do not necessarily equal the aggregate in terms of effects. Such is generally the case when the phenomenon under observation has synergistic effects or exhibits nonlinear characteristics. For example, two types of water pollutants taken separately may cause particular changes downstream. Taken together, however, they may behave synergistically; their combined effect may be more severe than their effects individually. Consequently, we cannot merely sum the damage each would cause independently to assess total effects.

Lack of consistent numerairies—the old problem of adding apples and oranges—creates difficulties. In developing measures which are susceptible to aggregation or disaggregation, consistent numerairies for disaggregates must be developed so aggregate measures will be meaningful. Otherwise, aggregate measures may be misleading or convey little useful information.

There are further difficulties. Important or relevant information may be hidden by aggregate measurement. Gross national product, for example, can be a relevant indicator, but it tells us nothing about the distribution of wealth, another important factor in the economy. However, such difficulties are not always resolved simply by employing appropriate disaggregation techniques.

Decision-makers and citizens are frequently overloaded with information. They may react by responding to only a selected subset of the information reaching them; this subset is usually determined by past conditioning rather than by a desire to search for new meaning. As a consequence, disaggregated information has little or no value when the individual may ignore most or all of it. Clearly, there are difficulties associated with the extremes of both aggregation and disaggregation. In devising social indicators, we need to select the appropriate level of aggregation to be reflected.

Finally, definition presents problems related to aggregation and disaggregation. For any given system, we can imagine a large number of relationships between macro- and microvariables. Consistency is a useful criterion for analyzing these relationships; a change in microvariables or in the microstructures will be reflected in a change in the macrovariable or in the macrostructure, and vice versa. If the macrovariable is an equally weighted sum of the microvariables, consistent interpretation is not difficult. An increase of 10 per cent in the unemployment rate (the macrovariable) indicates that .10X people are out of work, where X is the number of employable persons (the microvariables). With nonequally weighted sums such as price indices or most other aggregated indicators, a 10 per cent change in the macrovariable may yield information on overall effects but not on the individual changes in microvariables. In developing social indicators, it is crucial to define precisely the type of information desired and to design the indicator to yield that information. We must recognize, however, that designing an indicator for a particular use may conflict with alternative designs for other uses. The selection of aggregated or disaggregated indicators clearly needs to balance tradeoffs between perception, consistency, conceptual limits and multiple uses so that some degree of efficiency is realized.

Criteria for Aggregation and Disaggregation

In selecting the most desirable degree of aggregation, the central issues are *how much* and *where* to invest in raw information for indicator development. Information is costly and has many dimensions. Few would be willing to invest greatly for information generation. Therefore, in developing indicators, important decisions concern: 1) how much to invest in the generation of the social indicator; and 2) how to allocate that investment between definition, time, space and the type of measure to be developed. Two criteria can aid in this decision-making. The first concerns the accuracy and completeness of the measure involved; the second deals with efficiency in information generation.

In considering the accuracy and completeness of a measure, it is necessary to understand the precise nature of the phenomenon to be measured; it is then necessary to ascertain how well the proposed measure actually measures the phenomenon. Failure to understand this requirement can result in problems such as those illustrated by the federal government's use of dissolved oxygen content as one measure of determining whether a body of water is fit for swimming. The rationale for using dissolved oxygen as a measure stems from the general proposition that streams with high dissolved oxygen concentrations are generally not carrying heavy waste loads. However, the extent to which a stream can be used for swimming is not simply a function of overall waste loads, but rather a function of a very specific waste load, the number of pathogenic organisms in the stream. Dissolved oxygen concentrations indicate little or nothing about the presence or frequency of pathogenic organisms. As a result, reliance on the dissolved oxygen measurement alone could result in a stream being declared fit for contact when, in fact, it is not. In this instance the dissolved oxygen measurement is too aggregative a measure to be used. Clearly, the selected level of aggregation must be carefully tailored so the indicator can reflect accurately and fully the phenomenon being measured.

The second criterion concerns informational efficiency. Price is a relatively simple piece of information, yet prices convey vast amounts of information. The price system is perhaps the most efficient form of communication currently available. In devising indicators, alternative levels of aggregation need to be examined to see which provide information most efficiently.

INDICATORS IN PUBLIC POLICY

Social-environmental indicators can help man take intelligent, effective action on critical problems in his social and natural environments. We discuss next those characteristics of the political system affecting its capacity to serve as a feedback response and "feed forward" mechanism for maintaining the urban system in states compatible with human survival and fulfillment.

Obviously, the political system is only one control subsystem in the urban environment. A second, the economic system, exercises its control largely through marketing; environmental decisions, however, generally do not enter into market decisions. A third control system, culture, is the way a given human society characteristically behaves. Culture includes society's institutions, language, thought processes, value systems and artifacts. Culture regulates human behavior through the many social pressures on an individual to conform to the culture and through the individual's general inability to behave in ways other than those transmitted to him by preceding generations, his peers, or another culture. Other control systems in contemporary society are related to the management of information, such as advertising, public relations and mass media.

Among the various feedback-response mechanisms, the political system is of special interest in attempts to respond to critical problems and changes in the urban environment. It is the political control system which must confront the problems other control systems create or neglect.

Two types of governmental decisions significantly affect the social and natural environments: 1) decisions delineating or approving major projects and programs; and 2) policy decisions governing a large number of subsequent decisions. Under the terms of the National Environmental Protection Act of 1969, Section 203, both types of decisions at the federal level must be preceded by environmental impact statements. Such statements act as anticipatory performance statements for particular projects, policies or programs. They may also function as social-environmental indicators of forecast urban system states. After a project is finished or a program or policy has been instituted, it is then possible to measure actual consequences. Thus, we can have two types of indicators: 1) on anticipated and actual performance of a particular project, program or policy; and 2) on the anticipated and actual state of a social or environmental system as a result of all projects, programs, policies and other actions. The politi-

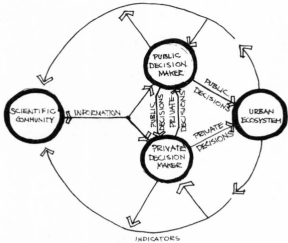

Figure 14

cal system is related to all the major systems taking action—communities or groups—and their pertinent attributes, decisions and actions (Fig. 14).

Uses of Indicators

All social groups in the political system use social-environmental indicators in two common ways: 1) to forecast consequences of alternate courses of action based on the results of previous actions; and 2) to determine action priorities based on indicators of various social and environmental policies. Scientific-professional communities (including planning agencies, universities, and research organizations) further use indicators: 3) to formulate and test theories (which will aid forecasting of consequences); and 4) to develop annual reports, plans, information systems, budgets, impact statements and other programs, planning and coordination. Public decision-makers (including legislative, executive and judicial officials at federal, state, regional, and local levels) use them: 5) to evaluate the quality of advice and recommendations from the scientific community); and 6) to justify decisions made against political pressures. Other communities and groups (including private businesses, special interest groups, lobbies, public interest groups and neighborhood associations) use indicators: 7) to determine the effect of proposed public decisions on particular goals and interests; and 8) to determine the effects of private business or consumer decisions on urban ecosystems.

The concept of an annual "state of the system" report is closely associated with the use of indicators. However, the question can be raised whether indicators should be annualized to conform to precedent or whether they should be timed to yield maximum information. The association between indicators and annual reports dates back at least to the Full Employment Act of 1946, which: established the responsibility of the federal government in managing the national economy; required the president to make an annual report to Congress on the state of the economy and on his proposed economic actions; provided a Council of Economic Advisors and staff to assist the president; and augmented the system of economic indicators to include current information on performance of the economic system.

Growing national concern about social conditions during the 1950s and 1960s led to a number of proposals on social indicators. In 1964, Raymond A. Bauer suggested an annual report to Congress from the president. In 1967, a group of senators led by Walter Mondale of Minnesota proposed a "Full Opportunity and Social Accounting Act" (S. 843), which required an annual social report from the president, to be studied by a Congressional Joint Committee. In January, 1969, in response to an earlier presidential request, the Secretary of Health, Education and Welfare, Wilbur Cohen, submitted a pilot study, "Toward a Social Report". Finally, in 1971, Congress passed the Mondale bill as "The Full

Opportunity and National Goals and Priorities Act" (S. 5).

Similarly, an annual report on the environment is now furnished by the president's Council on Environmental Quality. These pioneering efforts to develop a policy-response instrument using indicators to manage social and natural environmental systems have both recognized merits and limitations. During 1973, a number of bills were introduced or drafted in Congress to strengthen such accounting systems.

There are, however, significant differences between reporting on the economy and reporting on urban ecosystems. The history of economic statistics development significantly pre-dates that of statistics on social conditions in urban environments. Further, economic indicators are useful to the business community and have a strong, direct bearing on private decision-making. The policy goals of economic growth and stability tend to coincide with the goals of business; economic data collection has, accordingly, enjoyed strong support. In contrast, social and environmental data are more likely to prove embarrassing to private interests, particularly when coupled with goals of social equality. Support for social and environmental statistics has generally not been forthcoming from private interests.

The convergence of public economic development and private profit goals creates close ties between changes in the economic system and public and private responses to those changes. The social and environmental costs of private decisions, on the other hand, tend to be treated as external costs that fall outside the scope of private decision-making.

INDICATORS OF STABILITY AND RESILIENCY

There is a sizable amount of literature in many disciplines concerned with the structure and utility of and the interactions between stability, resiliency and diversity. Here, we bypass the confusion of much of that literature and operationally identify a set of definitions and functional associations helpful for indicating necessary and adequate indicators of stability and resiliency.

We have developed an identification scheme for the three top levels of organization in natural ecosystems (Fig. 15). These levels are critical response processes. The concept of stability is related to the consistency and/or predictability of systemic behaviors through time. At each level in the hierarchy or organization, the mechanisms and expressions of responses will vary. At the ecosystem level, we assume the trajectory of systemic development (succession) is bounded so as to converge at a common point (climax or terminal community) for any given environment. The consistent responses of landscapes and lakes to repeated perturbations such as fire, floods and agricultural disruptions are empirical evidence of the stability resulting from homeostatic mechanisms which operate at the ecosystem level. We assume that as long as the environment is not significantly altered, the pattern of ecological development will continue to repeat itself.

At the community level, response mechanisms operate to maintain the structure which characterizes any given stage of the successional trajectory. Regional stability results when biological populations compensate for changes in the environment through changes in internal states and environmental constraints. The natural community thus has the ability to withstand considerable perturbation if the disturbance does not exceed the capacity of compensatory response mechanisms.

The Dutch elm fungus, for example, has virtually eliminated the American elm from many eastern forests. Other species, however,

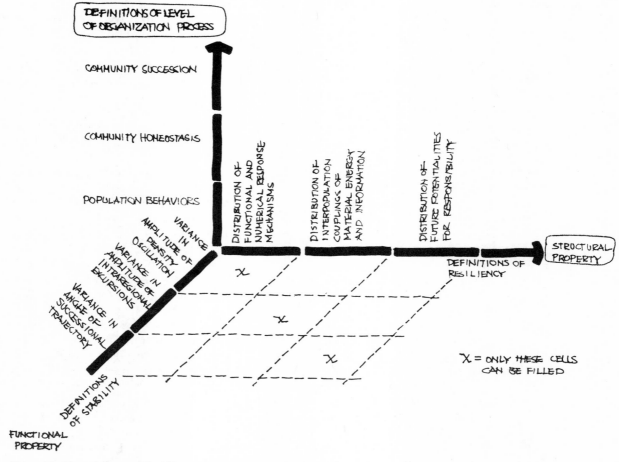

Figure 15 Natural Systems: Stability vs. Resiliency Indicators

have increased abundantly and have replaced the elms in many regions. The havoc created by the introduction of the sea lamprey and the alewife into the Great Lakes, on the other hand, exceeded the boundary conditions of the existing community, destroying it.

Individual populations oscillate in space and time for several reasons. Many problems associated with fish and game management, pest control, forestry, and agricultural production stem from the design and management technologies developed to control the temporal behavior of a particular population within man-imposed boundary conditions.

Stability is defined in terms of the statistical properties (variance) of behavioral responses at each level—ecosystem, community, and population. Resiliency, on the other hand, is defined in terms of the statistical properties (distribution) of the structural configurations at each level.

Utilization of Biological Indicators

In the course of time, organisms have developed a variety of unique sensory processes that gather, integrate and translate a wide array of environmental information. Whenever possible, these sensory processes should be used as the source of indicators for that subclass of information integrated by biological systems. A classical example is the use of canaries in coal mines as an inexpensive but sufficient indicator of carbon monoxide levels. Death of a canary is a patent indicator.

Biological indicators are most useful and dependable as devices providing early warning of short-term excursions into "trouble zones", particularly in urban environments. Active transport mechanisms enable biological organisms to "sense" low concentrations of chemicals efficiently. Physiological responses to input signals are rapid and can be easily monitored.

Considerable attention has been given to the use of indicator species to determine water quality; the attempts, however, are based on complex multispecies interactions, inhibiting any simple mechanistic interpretations. If mechanisms for interpretation can be well-defined, the use of sensory and response subsystems as indicators of short-term excursions into trouble zones may prove fruitful. Further, a range of response sensitivities could be obtained by selection of an appropriate array of organisms.

SYSTEM MODELS

A wide variety of quantitative and semi-quantitative techniques are used to analyze information about urban and other ecosystems. These techniques range from the conventional social science matrix models, such as input-output analyses (Leontief, 1951), through differential equation dynamic modeling approaches, as exemplified by the "Limits of Growth" study (Meadows and Meadows, 1972), to larger-scale analyses requiring massive computer-based technology (e.g., Hamilton, et al., 1969; Craven, et al., 1973). Each of these approaches has, to some degree, contributed successfully to our understanding of and solutions to problems of urban society. Each, however, has advantages and disadvantages in application.

We use the term 'model' in a broad sense here. It means any formal or semi-formal structure used to abstract aspects of the urban ecosystem for the purposes of analysis, problem-solving, or design. A conceptual structure need not be strictly quantitative in nature to be considered a model.

The rationale for using models in urban ecosystems stems from man's habit of abstracting aspects of reality when he is confronted with a perplexing situation. We can reason through only a few complex interactions at once; hence, modeling is useful as an attempt to group and symbolically restructure reality to deal with the behavior of important aspects within the limits of man's intellectual capacity.

In attempting to present a conceptual view of any system, the analyst must define the boundaries of the subsystems of particular concern and determine the level of generality or resolution at which he wishes to operate. He must also be aware that to remove parts from their context can be misleading and may result in seriously erroneous conclusions about the system's behavior. Indeed, considerable effort should be devoted to minimizing those dangers in systems analysis which may result subsequently in the application to the real system of a less-than-complete analog of reality. The dangers inherent in such applications are by no means unique to model building; they pervade all the empirical sciences. Nor do such dangers obviate the potential usefulness of the modeling approach. "The usefulness of mathematical models in the social sciences is pretty much an empirical issue that can be settled by experience with the construction and use of these models" (Hamilton, et al., 1969). Hamilton provides an excellent discussion of the technical usefulness and drawbacks of using mathematical models in the social sciences.

Modeling exercises to date have not been unqualified successes. As experienced modelers, therefore, we must approach the proposal of yet another model with considerable anxiety. The shortcomings of past models range from inadequacies of the conceptual base of the model to the lack of critical evaluation of results. Such shortcomings have many causes. Many models are produced specifically to achieve predetermined goals, yet their integrity goes unquestioned. This critical shortcoming is not necessarily the fault of participating modelers, but may be due to the complexity of the task they face. Further, unexpected assumptions may lie hidden in the modeling techniques which are chosen or which must be used.

Formal models must, therefore, be used with great caution. No tool can be more damagingly subverted to government advantage than expert testimony and scientific data. No opportunity for the advancement of civilization can be more tragically blunted than can the balanced awareness of our ecological dependence on nature and on one another. A model of these ecological relationships, involving extensive funding, extensive data collection, and potentially extensive implications for government policy, would provide considerable public relations leverage in the best of circumstances. In less than the best of circumstances, such information could be used coercively, possibly affecting man's relations with the natural environment quite negatively.

An Overview of the Model

Many monolithic models have been proposed to describe the urban system; they have been large and relatively undifferentiated. We call attention to the monolithic model to explicitly treat both the local area and the rest of the world. We hope, in this manner, to provide sufficient resolution to adequately discuss the details of the urban system while allowing realistic treatment of external factors, such as external resource management. If there were only a regional or larger scale model available, adequate detail might be precluded and, for example, minorities might be disenfranchised (an instance of reducing the diversity of the system being considered).

In general, the model we describe takes the local system, defined as the urban ecosystem,

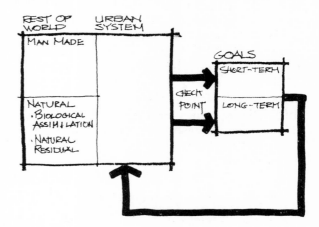

Figure 16

Processes: Interventions, Indicators, and Models 165

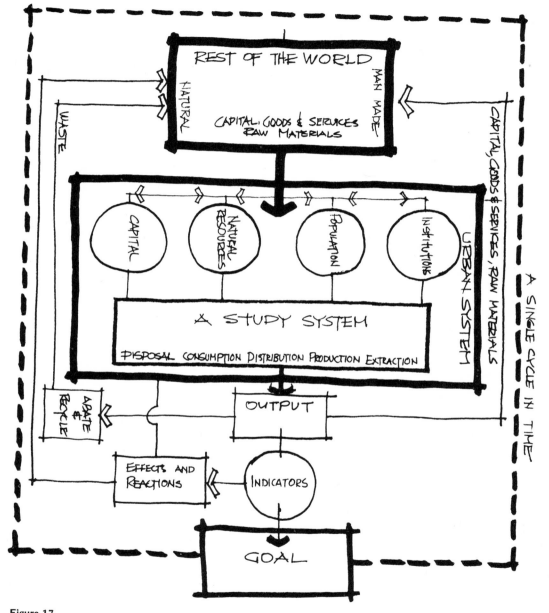

Figure 17

and places it in the rest of the world. The growth of the rest of the world, plus the local system, at a particular moment is related to the goal structure, both short and long-term; assuming that the desired ends of the system are met (i.e., in terms of the goals) the general model iterates in time from one period to the next (see Fig. 16).

A Conceptual Outline of the Model

Figure 17 is a preliminary attempt to expand the general overview presented in **Figure 16**. It attempts to tie together some of the structures suggested elsewhere in this volume, particularly those dealing with indicators of urban system health. **Figure 17** has the same general form as **Figure 16** and compares the growth of the urban ecosystem and its changes, both internal and external, to the rest of the world.

Reading the paradigm from left to right, we can see that the rest of the world, as structured at this stage of the model, consists of two parts: the man-made and the natural environment. The rest of the world, at a relatively high level of aggregation, is used primarily for two purposes: to serve as a relative sink for the production of goods and services; and to absorb goods, services, raw materials, capital, etc. It is the balancing arm and the driving force in the urban ecosystem. As we will show later, the growth and change forces stem from this sector of the model.

The local system, the focus of our discussion, has been explained in greater detail; we have already discussed goals, institutions, populations (both human and non-human), physical structure and function, and resource flows in earlier sections of this report. In our discussion of models, we emphasize the general links tying the various sectors of the urban system together.

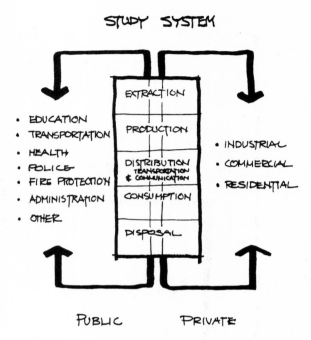

Figure 18

Figure 18 separates the public and private sectors of the urban system, including extraction, production, distribution, consumption, and disposal processes. The model indicates that the relative importance of public and/or private growth will help determine not only the rate of growth of the system under observation, but both the relative distribution of goods and services within it and the amenity level.

The public sector can be described in many ways; we have selected the simple budgetary approach and identify these elements: education, transportation, health, police, fire, administration, etc. The elements of the private sector are the industrial, commercial, and residential.

The interaction of capital, natural resources, population and institutions and the extraction, production, distribution, consumption and disposal processes each year produces output from the local system. The output is of two kinds: capital, goods, services, and raw materials; and wastes. Capital, goods, services, and raw materials flow from the local system to the outside world. At the same time, a similar bundle flows into the local system. Our measure of growth, then, will be in net terms: net material, net capital, etc.

The other form of output from the local system is the waste materials thrown off from urban processes. In our model diagram (Fig. 17), the wastes are directed into the rest of the world (the natural ecosystem) but have a shunt before arriving there. The shunt can be described as an abatement subsystem consisting of a change in the waste products; the wastes are either recycled into natural resources, or are generally abated. These processes have already been discussed in greater detail in terms of resource and residuals flows. If waste products are neither abated nor recycled in the human sector of the system, they must be assimilated by the natural system. To the extent that waste products exceed the capacity of the natural system to assimilate them at any time or in any locale, or if they exceed the capacity for abatement or recycling in the human sector, we have an accumulation of residual waste products.

The Cycles and Goals

Our model has thus far done little but describe in rather general terms a single cycle of the transformation of goods and energy into other forms. We need to relate the output of wastes to the natural limits of the system. When the ecologist talks in terms of the carrying capacity of the system, he means the relationship between the natural resources and the population of a species in a particular locale through time. The difference between the human animal and other species is that man's consumption rate is not constant, but is largely determined by factors other than his biological needs. Consequently, man's goals, indicators, quality of life, etc., in addition to natural resources, help define the carrying capacity of the urban ecosystem.

Indicators, in our paradigm, are viewed as the system's output, measured in terms of relative human values. Indicators can be of many types and, again, have been discussed in detail earlier. For our purposes here, we separate them into these general categories: economic, environmental (natural and man-made), social, and political. An indicator serves as a 'thermometer' inserted into the system to check its performance against system goals or the carrying capacity of the system. Goals of the urban ecosystem are of two types, short-term and long-term goals.

The model then iterates in the following fashion. After the local system produces and consumes over the course of a year, the indicators are used to discern whether at a particular time the system has reached its short-term goals. If the system has achieved those goals, the model moves forward; the reactions of the residuals or the like (these are the familiar feedback loops of any dynamic model) will cycle back into the model and be used to reset parameter coefficient values for the next iteration. If the indicator suggests that the short-term goal has been reached or exceeded, the model enters a second stage. In the second stage, the model checks the indicator value against the long-term goal of the system, to insure that long-term goals can be achieved at the current indicator level. To the extent that indicator values and long-term goals coincide, the model accepts the resultant feedbacks and iterates to the next level.

If indicator values and long-term goals do

not coincide, the model's reaction shunt is of a different type. The model must be rebalanced to move through the distribution process again to arrive at another level of output more compatible with the long-term goals of the system. Or, conversely, at this stage the long-term goals of the system can be redefined or revised. If the indicator check against the short-term goals reveals that indicator values and short-term goals do not coincide, the effects reaction section of the model would be used to recycle the model, by re-examining the system and re-checking output values, and would continue this process until the indicator values and the short-term goals coincided. If coincidence cannot be achieved, the system is either in decay (has reached the limits of its possible growth), or the goal statement is too great and will have to be reset by the local inhabitants of the system. Of course, the resetting of goal statements is possible even in the first iteration of the model if indicators and short-term goals do not match initially.

Perturbing the Model through Time

To change the model or to make this static description dynamic, we begin by making a major hypothesis (see Fig. 19). By definition, all the change components available to the total system at any particular moment are made up of local and extra-local change components. Because the relative size of the urban ecosystem in our model is unknown, we assume that its relative size, compared to the rest of the world, is quite small. Although the system may change at any moment locally, it will change not only itself, but the rest of the world as well; we assume that most of the change components with change levers will be accounted for in the extra-local system.

Instead of attempting to define relative

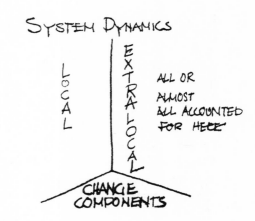

Figure 19

changes in technology, political beliefs, etc., we have listed the major change levers in **Figure 20**. These levers can be examined in several ways; one example may be useful to our purposes here.

Over the last several decades, a change in the major form of production has taken place in developed countries. When our culture was a subsistence culture, most forms of production were primary, i.e., they were extractive, agricultural, etc. In time, through the industrial revolution, the major form of production changed from primary to secondary, i.e., largely manufacturing. In recent years, large numbers of people have been employed in tertiary industry, i.e., service, etc. Now, the service industry is either being redefined or replaced by quarternary industry, or what many economists call the leisure industry group. Such a redefinition of changes through the economic arrangements sector of the model portends considerable system changes

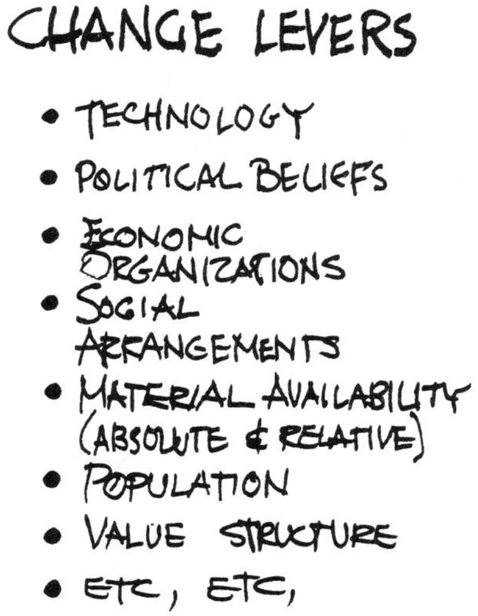

CHANGE LEVERS

- TECHNOLOGY
- POLITICAL BELIEFS
- ECONOMIC ORGANIZATIONS
- SOCIAL ARRANGEMENTS
- MATERIAL AVAILABILITY (ABSOLUTE & RELATIVE)
- POPULATION
- VALUE STRUCTURE
- ETC, ETC,

Figure 20

throughout the urban ecosystem. There may be changes in the labor mix, in types of production, in the location of various types of land use, in types of buildings constructed, etc. Consequently, the movement of the production technology change lever is pervasive throughout the system; it is not endemic to any local system, but has considerable effect in all local systems.

Our model indicates an iteration through a particular moment using parameters and coefficient values which are constant for that period of time. It is difficult to keep this data constant. Consequently, change components, both local and extra-local, can be viewed as potentially changing as the model goes from one time period to the next. Certainly, change levers would be the major factors in the readjustments.

Goals/Indicators

If goals and indicators are to be used in a model, they must eventually be described in the same terms (Fig. 21). If they are not, they cannot be compared. For example, some measure of consensus between a "quality of life" goal and the appropriate indicators is necessary. If the "quality of life" concept is viewed as a profile (i.e., left in disaggregated form, the economic, environmental, social and political indices), a number of independent measurements of the indicators and their goals are revealed. If any indicator is out of balance with the short- and long-term goals of the system, the model will be reiterated to adjust the system's output or to change the "quality

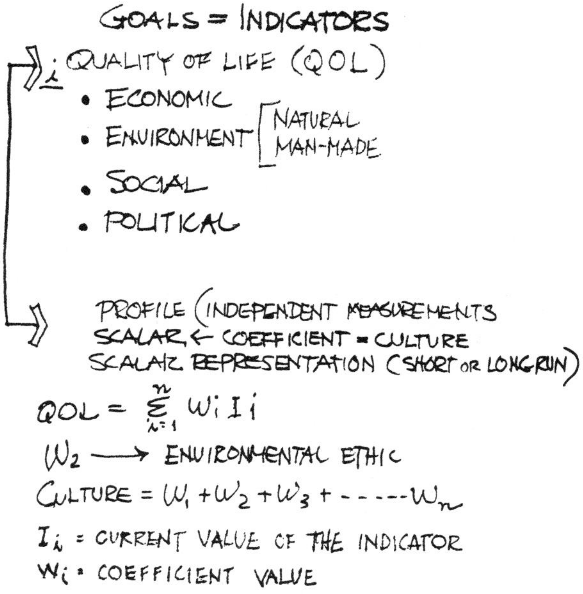

GOALS = INDICATORS

QUALITY OF LIFE (QOL)
- ECONOMIC
- ENVIRONMENT [NATURAL / MAN-MADE]
- SOCIAL
- POLITICAL

PROFILE (INDEPENDENT MEASUREMENTS
SCALAR ← COEFFICIENT = CULTURE
SCALAR REPRESENTATION (SHORT OR LONG RUN)

$$QOL = \sum_{i=1}^{n} W_i I_i$$

$W_2 \longrightarrow$ ENVIRONMENTAL ETHIC

$CULTURE = W_1 + W_2 + W_3 + \cdots W_n$

I_i = CURRENT VALUE OF THE INDICATOR
W_i = COEFFICIENT VALUE

Figure 21

of life" vector. If, on the other hand, the components of the profile are combined to form a scalar, the coefficients can only be used to define culture, either in the short- or long-term sense. For example, the following simple description of the quality of life,

$$\sum_{i=1}^{n} W_i I_i$$

gives us a formula for computing such a scalar. The individual coefficients (W_1, W_2, ... W_n) can be described as a representation of a society's culture at a particular moment and can be reconstituted to determine the society's culture in the long-run. The relative value of the coefficient representing the natural environment in relation to the values of the rest of the coefficients can be used to determine the society's environmental ethic. Unless the environmental coefficient has a long-term value sufficiently high enough to achieve some of society's goals, a shift in culture will reduce the environmental ethic to a short-term phenomenon.

Types of Models

We have defined a modeling approach that we feel is useful for describing the urban ecosystem. It may be implemented for simulation in many ways, depending on the particular definition of the system and what one wishes to learn from it. Then we took our general paradigm and described how we would model such a system.

The model itself should not be totally analytical, but more appropriately a man-machine model; during each time interval of the simulation, the indicator should be checked to ascertain that the outputs are realistic and that the goal statements are still those desired (see Fig. 22).

Several types of models can be combined for this approach (see Fig. 23). Internally

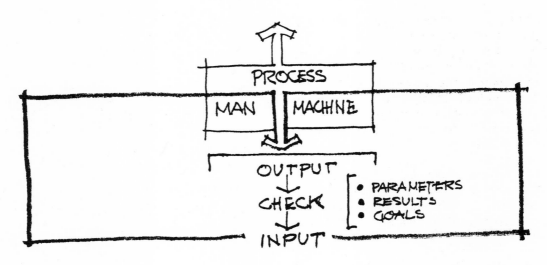

Figure 22

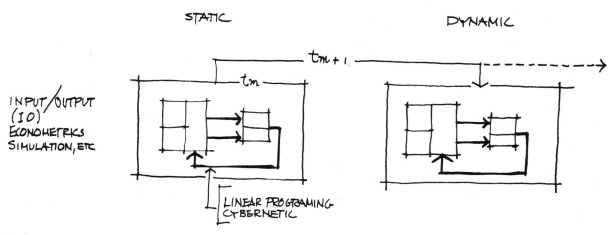

Figure 23 Simulation Model

within a single time cycle, the model is a linear-programming type or cybernetic model which, instead of producing the optimal response, attempts to produce one which is simply adequate. Any time the indicators of the goals are within their specified limits, the model is allowed to advance to the next time period. If not, adjustments are required within the system and can be made by using any static model (e.g., Input/output, econometric, simulation, etc.). Performance of the model through time can be implemented with any dynamic-simulation techniques which allow man-machine interaction.

The following brief descriptions of models that have been or are being used experimentally provide further understanding of simulation techniques. Each approach differs from the others in one or more fundamental ways, among them:

1. **Spatial resolution**—is the model to assign information to one, two or three-dimensional geography?

2. **Static vs. dynamic characteristics**—is the structure of the model based upon deterministic mathematics which are constant in time (e.g., algebraic or time independent matrix equations), or upon dynamic feedback relationships which modify model results through time (in which case the model is said to be "behavioral" or "predictive")?

3. **Feedback control**—does the model include a dynamic controller so desired values of the variables through time (the "goal") can be specified; does the model attempt to approximate these values by self-modification of parameters?

These characteristics can appear in various combinations in any given urban system model, such as those which follow:

1. **"Data Display" and Manipulation Techniques.** This approach includes a range of methodologies, from a data file to a computer-based information storage and retrieval network. In the data file, output would be obtained manually, perhaps in tabular or graphic form; little processing or manipulation of the data takes place in such a system.

In the computer-based information network, extensive processing is possible; output can be automatically produced in tabular and graphic forms; this system can be spatially detailed, but has little predictive capability and is not dynamic.

2. "Expert Team" Analysis. This approach usually involves several specialists jointly studying particular urban problems or a particular urban system. The team can employ a purely qualitative approach, using sketches, written descriptions and diagrams; or it can use mathematics, possibly with computer support. The results of such a study represent the consensus of the participants. The approach can be predictive and quite spatially detailed.

3. "Static Matrix" Technique. This approach is quantitative and heavily dependent on data. The matrix operations usually require computer support; material is abstracted from more comprehensive files or personal interviews and record-searching. The static matrix approach can handle many variables, but its predictive capability is limited.

4. "Small-Scale Differential Equation" Dynamic Model. This approach describes an urban system or subsystem by a set of differential equations. The equations, either linear or nonlinear, are solved on a computer using standard numerical techniques. Feedback loops give the model a "dynamic" behavior and make it predictive. This model does not have much spatial resolution and its predictions are usually coarse or highly aggregated.

5. "Large-Scale Differential Equation" Dynamic Model. This can be simply a larger version of the above approach, but may include a capacity for spatial resolution as well. If spatial resolution is required, the approach requires considerable data.

6. "Control System" Imposed Upon the Differential Equation Approaches. This approach utilizes a control module in combination with the dynamic differential equation models. The control module adjusts the parameters of the differential equations during the model's execution to obtain a specified time trajectory. The technique is well-established in the aerospace industry and electrical engineering. Work by Koenig (Koenig, 1973; Koenig, et al., 1972) suggests possibilities for the use of this approach in urban ecosystem analysis.

7. "Static Matrix/Differential Equation" Model. This approach would utilize a static model in tandem with a dynamic differential equation model. The dynamic model would provide predictive capability; the static model would spatially distribute the values throughout the urban ecosystem. Apparently this model has not been applied to the urban ecosystem, although it appears to be appropriate.

8. "Control System with Static Differential Equation Model". This method would combine the static matrix/differential equation approach with a control module. Potentially, it would be spatial in resolution and predictive, with optimal control capability. This approach appears beyond current urban ecosystem modeling capabilities.

Criteria for Model Evaluation

It is difficult to establish uniform criteria for evaluating models. Our framework is not comprehensive, and simply attempts to detail most pertinent considerations. This framework will serve as a method for making interim judgments about models and as a base from which an improved framework might be developed.

A. Model capabilities and objectives. 1) Time predictivity—is the model temporally predictive? Is it dymanic, rather than representative of a system at a single moment? 2) Spatial discrimination—does the model make distinctions between different regions or points in space, or is it descriptive of a single homogeneous area? 3) Regional specificity—is the model intended for use in a particular region or type of region (e.g., coastline, highly developed area, developing area, etc.), or can it be applied equally well in any area? 4) Problem oriented—does the model answer specific questions (e.g., how many people can San Diego serve without degrading air quality?), or does it mainly show general trends (e.g., as technology increases, how does the quality of life change?)? 5) Multisector capabilities—does the model have explicit subsections for different sectors of society, or does it use an aggregated approach which lumps many sectors?

Additionally, what are the model's specific capabilities with regard to: 6) population size, distribution, makeup; 7) resource use, availability, limits; 8) transportation networks; 9) public utilities; 10) private income; 11) standard of living/quality of life; 12) legal and institutional structures; and 13) alternative value systems or social goals?

B. Model structure and framework. 1) Data Intensive—does the model require a large number of concrete statistics, or are general aggregated facts and intuitive mechanisms sufficient? 2) Parameters identifiable—are the parameters, in general, equivalent to standard measurable indicators, or is model-specific massaging of data required? 3) Highly aggregated—are the model variables highly aggregated with low resolution, or disaggregated with high resolution? 4) Explicit Hypothesis—does the modeling technique have or require explicit statement of particular or controversial hypotheses for its mechanisms, or does it depend on widely accepted and standard techniques?

C. Development logistics. 1) Cost—is a high cost (more than $500,000) associated with the model's development or use? 2) Technology—is the technique at the frontiers of what technology allows, or are the basic tools widely available and cheap? 3) Communicability—is the method of the model readily communicated to users, or is it extremely detailed or complex, requiring much study and experimentation before it can be used even by professionals? 4) Implementation—has the technique been successfully used to influence legislation, plan public or private investment or alter the structure of society in some way, or is there strong potential for such use?

SUMMARY

If man is to shape the urban ecosystem to his needs, he must be able to intervene wisely and effectively into the physical structure and material and energy flow systems of the city. Intervention requires that we be able to adequately measure urban system performance and compare those measurements with specified goals. Modeling techniques can be used to abstract aspects of the urban ecosystem for the purposes of analysis, problem-solving, or design.

We have discussed here the need for identifying appropriate alternatives for improvement of urban subsystems, and have indicated various tradeoffs involved in choosing one kind of intervention over another. We have also discussed the natural and human constraints which limit our capacity to intervene effectively; these constraints include technology, resource availability, and the limited capacity of the environment to assimilate wastes, etc. Decisions about action to improve urban system performance must recognize such constraints and necessary tradeoffs.

Social-environmental indicators of urban system health allow us to measure system performance in particular areas and to check that performance against specified goals. The indicators we develop should inform us when the system is moving into one or more "trouble zones" of social, economic, political or ecological irreversibility. Indicators are selected on the basis of the information they will provide for evaluation of progress toward man's goals in the urban system.

Modeling the system allows us to group and symbolically restructure reality so that we can deal with the behavior of important aspects within the limits of our intellectual capacities. Modeling exercises have been only qualified successes to date. The model we propose allows us to view the urban system both as a static system at a particular moment, and as a dynamic system as it moves through time. Indicators are necessary for evaluating the system's progress; feedback loops in the dymanic model allow it to recheck the system when short- and long-term goals have not been met, and to restrict processes or redefine goals as necessary.

We must emphasize that great caution is to be taken in the use of formal models. No tool can be more damagingly subverted to special interest purposes. Modeling can too easily be used as public relations device, or worse, as a weapon for coercion.

Finally, we discuss techniques which have been or may be used in modeling the urban system, and list an initial set of criteria against which proposed models can be checked.

By intervening wisely in the urban system, using carefully selected system indicators and carefully developed models, we can better understand and more effectively shape the urban system to meet human and environmental needs.

RECOMMENDATIONS

Policy Recommendations

III-1. We recommend that the national government encourage comprehensive urban planning by coordination of research, data gathering, development of innovations, and evaluation and that this information be exchanged through a central data bank. In this way, a city may benefit from information gained in studies of other cities.

III-2. We recommend that urban planning programs recognize fully the interrelationships among urban physical structure, land use patterns, social behavior, energy utilization, and the biological needs of residents.

III-3. We recommend that government at all levels develop and declare policies that explicitly recognize the finite capacity of the environment to supply resources and accept residuals. Further, when policy-makers are developing strategies to reduce resource depletion and residuals discharge, they should consider: the complete system of energy and material flow, (including resource extraction, product generation, consumption, and residuals discharge); the decision-making socio-economic-political system; and the implicit costs associated with intervening (or not intervening).

III-4. We recommend that, in the implementation of policies, the total costs and total benefits of alternative actions be examined; all resource-residual effects and environmental, economic and social consequences should be weighed for each alternative action.

III-5. We recommend that, in the selection of indicators of urban ecosystem health, highest priority be given to providing methods for detecting irreversible losses of flexibility in the system.

III-6. We recommend that future urban ecosystem modeling projects include mechanisms to encourage interdisciplinary communications; wide-spread dissemination of techniques and results will be necessary for this purpose.

III-7. We recommend that support be provided for development of the technology needed to manage waste energy.

Recommendations for Action

III-8. We recommend that governments reduce energy and materials flow by improving the product specifications for their own structures and equipment and by more efficient operating practices.

III-9. We recommend the establishment of minimum efficiency standards for furnaces, air conditioners, and appliances; we further recommend the inclusion of energy consumption information on name plates and price tags and in advertisements for appliances; and we further recommend use of fiscal, legal, and psychological incentives to discourage unnecessary packaging, to reduce the production of highly polluting materials, and to encourage the manufacture of longer-lived products.

III-10. We recommend that the following principles underlie the selection and development of indicators of urban system performance:
1) The sensory processes which organisms have developed should be used for that subclass of information involving biological systems.
2) The level of aggregation should be tailored so that indicators can reflect accurately and fully the phenomenon being measured.
3) The units of measurement for disaggregates should be consistent so that aggregate measurements will be meaningful.
4) The selection of aggregated and disaggregated indicators of urban system health should balance the tradeoffs between perception, consistency, conceptual limits and multiple uses so that some degree of efficiency is realized in indicator use.

III-11. We recommend that the United Nations establish a formal, integrated, continuing program for the collection and annual publication of socio-economic-environmental indices. This program could: stimulate indicator research and utilization; minimize governmental tampering with indicator data; and promote standardization in data collection.

III-12. We recommend initiation of small-scale research projects in urban areas to explore the possibility of integrated data collection and of applied indicator research. The use of secondary data sources would reduce costs. These projects should help to unify presently fragmented indicator research efforts.

III-13. We recommend that the information stored in data banks be expressed in understandable language, that the system be accessible and that tampering be prevented by insuring that only evaluations of the data and not the data themselves can be revised.

III-14. We recommend, for large-scale urban modeling, expansion of our understanding of values to provide a common measure for both traditional wealth and commonly-owned environmental resources and attributes. Without such an understanding, decision-making involving tradeoffs between the two are impossible. The testing of such a common measure in modeling systems seems to be a prerequisite to its incorporation into actual situations.

Research Recommendations

III-15. We recommend research on ways to improve environmental impact analysis and other evaluation processes for urban systems.

III-16. We recommend research to improve methods for evaluating alternative courses of action in the urban system.

III-17. We recommend research to identify opportunities for and constraints on the improvement of urban configuration. These opportunities and constraints should be correlated to preferred or potential land uses.

III-18. We recommend increased research on tradeoff techniques for making choices between alternative land use proposals. Improved techniques should assess consequences of action relative to human needs and ecological soundness.

III-19. We recommend study of the spatial distribution and effects of uncontrolled energy dissipation in urban ecosystems.

III-20. We recommend research on alternative technologies to support urban activities such as building construction, transportation and communications.

III-21. We recommend research to determine the effects of scale on energy flow, materials cycling and structural parameters and to identify the nature and scale of unintended distribution of materials by various transport systems.

III-22. We recommend research on the ways that price, utility and advertising influence the values an individual may place on products and on the environment.

III-23. We recommend research to determine whether regulatory or market incentives are more appropriate tools for intervening in the urban materials and energy system.

III-24. We recommend investigation of the potential usefulness and possible detrimental effects of a pollution tax as a tool in controlling discharge of pollutants.

III-25. We recommend research on the design components of a systematic set of urban ecological indicators.

III-26. We recommend substantial research efforts on: the definition of and interactions between the flows of material, energy, and information through large urban ecosystems; and on the adaptability, stability, and resiliency of urban material, energy, and information systems.

III-27. We recommend that critical flows in the urban system be defined, quantified, and projected into the future and that fundamental limits or "choke points" be sought.

III-28. We recommend that societies from the primitive to the post-industrial be studied to examine interactions among the following factors: a) system structure; b) values; c) information generation, processing, transmission, retention, and projection into subsequent generations; d) system controls; e) indicators or sensors utilized by the control subsystem and by outside investigators.

III-29. We recommend further study of the precise role of energy as a causal factor of or concomitant factor to the development and maintenance of the structure of urban systems. The role which energy could play as a key index to human values and goals requires particular attention.

REFERENCES

Abelson, Philip H. 1972. Energy conservation. Science 178:355.

Alonso, W. 1970. The economics of urban size. Working paper 138, Institute of Urban and Regional Development, University of California, Berkeley.

Anon. 1971. Can building codes help protect the environment? Progressive Architecture 52(10): 125.

Anon. 1973. Currents. Environmental Science and Technology 7(5):391.

Anon. 1973. Energy: save a watt. Newsweek 81(28 May 73):90,92.

Anway, J. D., E. G. Brittain, H. W. Hunt, G. S. Innis, W. J. Barton, C. F. Rodell, and R. H. Sauer. 1972. ELM: Version 1.0, Technical Report No. 156, Natural Resource Ecology Lab., Colorado State University, Ft. Collins.

Arrow, K. J. 1963. Social choice and individual values. 2nd edition. Yale University Press, New Haven.

Ayres, Robert U., and Allen V. Kneese. 1971. Economics and ecological effects of a stationary economy. Annual Review of Ecology and Systematics 2:1–22.

Ball, John. 1973. Insulation '73—a viable industry? Roofing/Siding/Insulation (January, 1973).

Berry, B. J. 1964. Cities as systems within systems of cities. Pages 116–137 in W. Alonso and J. Friedman, eds. Regional development and planning. MIT Press, Cambridge, Mass.

Berry, Brian J. L. 1961. City size distributions and economic development. Economic Development and Cultural Change 9:573–588.

Booth, John. 1971. Cost reduction in Novia Scotia schools through improved use of records. Nova Scotia Technical College, School of Architecture Report No. 10, March 1971.

Bower, B. T., O. G. Lof, and W. M. Hearon. 1973. Residuals in the manufacture of paper. Proceedings of American Society of Civil Engineers 99:Proc. Paper 9543.

Bryson, R. A. 1973. Environmental characterization management and planning. Third Six-Month Progress Report of NSF, RANN Division, from Institute of Environmental Studies, University of Wisconsin-Madison, January 1973.

Chapman, Duane, Timothy Tyrrell, and Timothy Mount. 1972. Electricity demand growth and the energy crisis. Science 178:703–708.

Clark, Colin. 1945. The economic functions of a city in relation to its size. Econometrica 13:97–113.

Clavel, P., et al. 1969. Alternative organizational models for district development. Mimeo. Regional Development Studies, Department of Rural Sociology, Cornell University, Ithaca, New York.

Commoner, Barry, Michael Corr, P. J. Stamler. 1971. The cause of pollution. Environment 13(3):2–9.

Commoner, Barry, 1971. The closing circle. The Viking Press, New York.

Cook, Earl. 1972. Energy sources for the future. The Futurist 6(4):142–150.

Corr, Michael, and Dan MacLeod. 1972. Getting it together. Environment 14(9):2–9, 45.

Craven, C. W., Jr., F. P. Baxter, C. R. Meyers, Jr., R. J. Olsen, C. R. Schuller, A. H. Volker. 1973. Regional environmental systems analysis progress report. June 1971–June 1972, Oak Ridge National Laboratory Report, ORM-NSF-EP-12.

Dorney, A. S., and G. B. Priddle. 1973. Inventory, analysis, and assessment of urban environmental quality. Faculty of Environmental Studies, University of Waterloo, March 1973.

Erlich, P. R., and J. P. Heldren; Commoner, Barry. 1972. The closing circle: criticism and reply. Environment 14(3):23–26, 31–52.

Forrester, J. W. 1969. Urban dynamics. MIT Press, Cambridge, Mass.

George-Roegen, N. 1971. Entropy, law and the economic process. Harvard University Press, Cambridge, Mass.

Grimmer, D. P. and K. Luszczynski. 1972. Lost power. Environment 14(3):14–22, 56.

Hamel, B. B., and H. L. Brown. 1972. Utilization analysis of energy systems. Proc. Effective Energy Utilization Symposium, Drexel University, June 1972.

Hamilton, H. R., S. E. Goldstone, J. W. Milliman, A. L. Pugh III, E. B. Roberts, A. Zellner. 1969. Systems simulation for regional analysis—an application to river-basin planning. MIT Press, Cambridge, Mass.

Hammond, Allen L. 1972. Conservation of energy: the potential for more efficient use. Science 178:1079–1081.

Hammond, Allen L. 1972a. Energy needs: projected demands and how to reduce them. Science 178:1186–1188.

Hannon, Bruce M. 1972. Bottles, cans, energy. Environment 14(2):11–21.

Healy, T. J., and D. T. Dick. 1972. Energy costs of an electric mass transit system. Status Report No. 1, University of Santa Clara, July 1972.

Hett, Joan M. 1971. A dynamic analysis of age in sugar maple seedlings. Ecology 52(6):1071–1074.

Hirst, Eric, and John C. Moyers. 1973. Efficiency of energy use in the United States. Science 179:1299–1304.

Hirst, Eric. 1973. Energy-intensiveness of transportation. Proceedings of American Society of Civil Engineers 99:TEI, February 1973.

Koenig, H. E. 1973. Human ecosystem design and management—a sociocybernetic approach. Manuscript prepared under NSF Grant GI-20. Department of Electrical Engineering, Michigan State University, East Lansing.

Koenig, H. E., W. E. Cooper, and J. M. Gulvey. 1972. Engineering for ecological, sociological and economic compatibility. IEEE Transactions on Systems, Man and Cybernetics, July 1972.

Lampard, E. 1955. History of cities in economically advanced areas. Economic Development and Cultural Change 3, January 1955.

League of Women Voters of the United States. 1972. Recycle. League of Women Voters Education Fund, Pub. No. 132.

Leontief, V. W. 1951. The structure of the American economy: 1919–1939. 2nd edition. Oxford University Press, New York.

Lincoln, G. A. 1973. Energy conservation. Science 180:155–162.

McPherson, M. D. 1971. Feasibility of the metropolitan water intelligence system concept. ASCE Urban Water Resources Research Program, Technical Memorandum No. 15.

McPherson, M. D. 1972. Hydrologic effects of urbanization in the United States. ASCE Urban Water Resources Research Program, Technical Memorandum No. 17.

Makhijani, A. B., and A. J. Lichtenberg. 1973. An assessment of residential energy utilization in the USA. College of Engineering, University of California, Berkeley, Electronics Research Laboratory Memorandum No. ERL-M370.

Makhijani, A. B., and A. J. Lichtenberg. 1972. Energy and well-being. Environment 14(5):10–18.

Meadows, Dennis L., and Donella H. Meadows. 1972. The limits to growth. Wright-Allen Press, Cambridge, Mass.

Mera, K. 1970. On the concentration of urbanization and economic efficiency. Economics Department, Working Paper No. 74, International Bank for Reconstruction and Development, Washington, D.C. Monograph.

Morowitz, H. J. 1968. Energy flow in biology. Academic Press, New York.

Odum, Eugene P. 1970. Optimum population and environment: a Georgia microcosm. Current History 58:355–359.

Odum, H. T. 1971. Environment, power and society. Wiley-Interscience, New York.

Office of Emergency Preparedness. 1972. Potential for energy conservation. Executive Office of the President, October, 1972.

Parks, Gary A. 1972. Civil engineering—a key role in the energy crisis? Civil Engineering 42:85–86.

Passonneau, J. R. 1972. Urban environment and urban transportation: a case for environmental investments. Presented to AIP National Convention, October 1972.

Penny, A., and P. Manning, 1973. Building environmental studies in Nova Scotia. Submission to RAIC from Nova Scotia Association of Architects, February 1973.

Perelman, Michael J. 1972. Farming with petroleum. Environment 14(8):8–13.

Redfield, R. 1955. The cultural role of cities. Economic Development and Cultural Change 3, January 1955.

Rice, R. A. 1973. System energy and future transportation. Technology Review, January 1973.

Royal Architectural Institute of Canada. 1973. Brief to the Science Council of Canada Concerning Energy Conservation, April 1973.

Russell, C. J., and W. O. Spofford, Jr. 1972. A quantitative framework for residuals management decisions. *In* A. V. Kneese and B. T. Bower, eds. Environmental quality analysis: theory and method in the social sciences. The John Hopkins Press, Baltimore, Maryland.

Spofford, W. O. Jr., C. J. Russell, and R. A. Kelly. 1972. Operational problems in large scale residuals management models. Presented at Universities-National Bureau Committee Conference on Economics of the Environment, University of Chicago, November 1972.

Spofford, Walter O., Jr. 1971. Solid residuals management: some economic considerations. Natural Resources Journal 11:561.

Stein, Richard G. 1972. A matter of design. Environment 14(8):16–20, 25–29.

Stein, R. G. 1971. Architecture and energy. Preprint AAS Philadelphia Meeting, December 1971.

Van den Hazel, B. J. An ecologist looks at modern school design. Building Cons. Council, Ontario.

Vancouver regional simulation study. 1971–72. Resource Science Centre, University of British Columbia, Second Year Report, 1971–1972.

Vapnarsky, Cesar A. 1969. On rank-size distributions in cities: an ecological approach. Economic Development and Cultural Change. 17(4):584–595.

Vining, R. 1955. A description of certain spatial aspects of an economic system. Economic Development and Cultural Change 3, 1955.

Woldenber, Michael J. and Brian J. L. Berry. 1967. Rivers and central places: analogous systems? Journal of Regional Science 7(2):129–139.

Wolman, Abel. 1965. The metabolism of cities. Scientific America 213(3):178–190.

Young, Frank W. 1966. A proposal for cooperative cross-cultural research on intervillage systems. Human Organization 25:46–50.

Zipf, G. K. 1949. Human behavior and the principle of least effort. Addison-Wesley, New York.

SECTION FOUR:
Case Studies

An examination of selected case studies in the context of urban systems may allow us to capitalize on existing practical expertise in the management of the urban environment. Many consulting firms in the United States and Canada are now marketing their expertise in urban affairs. These consultants have backgrounds ranging from the traditional fields of architecture, landscape architecture, forestry, natural science, geological science, to the newer fields of environmental engineering and applied ecology. These practicing professionals often take enthusiastic and innovative approaches to urban problems and it seems appropriate to assess some of them. In many cases, consulting firms are able to do applied environmental management studies not feasible in the academic sphere.

An examination of case studies may also encourage closer ties between urban environmental "practitioners" and academic scholars. If academic research is designed with practitioners in mind, suggested programs can be implemented more readily, eliminating the five- to ten-year lag that results when research results are directed to scholarly journals and require reinterpretation and translation.

The six case studies we review in our discussion were selected according to most of the following criteria: 1) they are not regional or rural studies but are centered in urban or heavily settled areas; 2) several months or years have elapsed between the completion of the studies and our evaluation, aiding assessment of the studies; 3) each study demonstrates a contrasting approach; together they reflect a wide spectrum of methodologies and costs; 4) for all the studies, there are consultants' reports or some published documentation; 5) the professionals involved were available to contribute first hand to our discussion.

The studies we have selected are representative, but should not be considered inclusive. Consideration of other studies may have been

Table 1 Study areas

	Butte, Montana	Minneapolis, Minnesota	Waterloo, Ontario	Halifax, Nova Scotia	Knoxville, Tennessee	Norman, Oklahoma (region)
Population	40,000	1,900,00	40,000	225,000	701,000	1,000,000
Physical Size	12 sq. miles (estimated)	3,000 sq. miles	12 sq. miles	6,484 acres	6,700 sq. miles	50,000 sq. miles (includes support area)
Study Unit Map Scale	N/A	1:62,960	1/4 acre cells	1 hectare cells	170 acre cells	N/A
Methodology	Short-term design animation by small study team	Resource analysis for planning guidelines	Physical/amenity feature mapping	Resource analysis, gaming simulation, design schema	Model simulation	System simulation
Disciplines Represented in the Study	Architecture, Ecology, Geology	Landscape Architecture	Geology, Planning, Ecology	Architecture, Landscape Architecture, Engineering, Sociology	Systems Engineering	Economics, Sociology, Biology, Political Science, Mathematics, Meteorology, Agriculture
Public Participation Potential	High	Medium	High	Medium	Low	Low
Applicability to Non-Industrial Countries	High	Medium	Medium/High	Medium/High (non-computer part)	Low	Low
Duration of Study	3 days	1 year	3 months	18 months	5 years	4 yrs. study
Man Months	1½	42	8–10	30	1,620 est.	100
Costs	$2,500	$75,000	$20,000	$30,000	$2,000,000	Unknown but close to Knoxville study

equally valuable. What we intend here, simply, is a review of the "state of the art" in analyzing urban systems and an examination of implementation procedures on the basis of existing professional knowledge and expertise.

Four of the six studies we discuss were made in the United States: Butte, Montana; Minneapolis, Minnesota; Knoxville, Tennessee; and Norman, Oklahoma. Two took place in Canada: Halifax, Nova Scotia; and Waterloo, Ontario. The methodologies of the studies range from short-term design animation to systems simulation. Manpower requirements, scale of detail, and costs also varied widely (see Table 1). The Norman, Oklahoma, case is more properly considered a study approach since the project has not been conducted; the other studies have already been implemented, at least in part.

Those who presented material included in our discussion were: Royce La Nier on Butte; Robert Einsweiler and James Veltman on Minneapolis; Robert Dorney on Waterloo; Peter Jacobs on Halifax; Peter Craven on Knoxville; and Amos Eddy on Norman.

HIERARCHICAL ENVIRONMENTAL ANALYSIS

If urban ecosystems are to be understood and managed in harmony with natural processes, the scale of detail necessary for analyzing them must be made clear. Many resources such as surface or ground water, for example, cross political boundaries. Any analytical method or system to manage such resources must be able to reflect overlapping jurisdictions. Accordingly, a hierarchy of information levels is needed, ranging from individual natural resource grid areas from three to 100 acres for the rural and urban fringes, to grid areas as small as one quarter of an acre (the size of a suburban lot) for study units within the city. Likewise, the man-power needs, costs, and study approaches depend on the size of the area and the level of detail to be studied.

For example, those factors affecting the siting of a large electric power plant or a major seaport facility in a metropolitan area may be different from those affecting the siting of a small commercial shopping center. Similarly, the scale of data require to analyze the effect of major land use shifts on the primary productivity of forests or grasslands or on large-scale weather patterns is different from the scale required to consider the effect that paving central city parks will have on local bird populations or on central city temperatures.

Our case study examination, dealing with hierarchical analyses at three levels (state or regional, county or large metropolitan, and urban; see Table 2), encourages explicit consideration of properties at each level and helps maintain an appropriate balance between generality and detail. Further, the establishment of a data source on a regional or large urban scale helps eliminate repetitious search for and storage of data.

Our review of case studies at the state, county and urban levels addresses environmental issues at each level (Table 3). Although we have not reviewed any case studies covering the national-international scale, we believe that consideration of many urban ecosystem-related problems will require appropriate integration at this level of detail or generality as well. Hence, consideration of the national-international scale is included here for conceptual completeness.

Level I: The International and National Scale

The need for an environmental information and referral service on a world-wide scale may

Table 2
Organization of a Hierarchical System of Environmental Information

INTERNATIONAL/NATIONAL

LEVEL I	*Systems Capability*	Land use, technology assessment
	Data Bank	Natural resource reconnaissance satellite weather information
	Staff Background	Multi-disciplinary

STATE/MULTI-COUNTY AREA

LEVEL II	*Systems Capability*	Socio-economic, socio-political Land use Ecological Geographical
	Data Bank	Census data Geology, soils, hydrology, forest lands, fisheries and wildlife statistics Land capability Cultural and historical land use trends
	Staff background (5-10 people)	Economics, systems analysis Mathematics, ecology, sociology, engineering, hydrology

COUNTY/METROPOLITAN AREA

LEVEL III	*Data Bank*	Land capability and suitability scales of 1:50,000 or 1:25,000 Soils, geography, hydrology, forests, wildlife and recreation statistics Present land use Cultural and historical land use trends
	Staff Background	Environmental extension specialist: science background, design background

LEVEL IV

	CITY 100,000 OR MORE		CITY LESS THAN 100,000
Data Bank	Amenity resources Water quality Noise Vegetation Open space Ground water	*Data Bank*	Same as city of 100,000 or more
Staff	Wildlife biologist, environmental scientist or ecologist and environmental designer	*Staff*	Part-time or professional retainer

soon be met by a United Nations' Environment Program (UNEP) project which grew out of the Stockholm Conference on the Human Environment. Another UNEP program, "Earth Watch", will develop a system for monitoring environmental quality

Table 3
Environmental Issues Addressed at Each Level in the Hierarchical System

LEVEL I: Large-scale trends in land use; impact of chemical products on land and water resources; impact of new technology on land and water resources.

LEVEL II: Impact of industrial location on air, water, waste disposal; impact of new town locations; impact of growth centers; impact of locating major facilities like harbors, roads and airports.

LEVEL III: Land use zoning; performance standards; subdivision design approval; growth options or directions for urban expansion; site selection and operation of solid and liquid waste disposal operations and local citizen participation.

LEVEL IV: Zoning for environmental effects; setting standards on noise levels, open sapce and water quality; park location; plant and animal diversity and management.

throughout the biosphere. A comparative analysis of urban ecosystems, many of which exert significant influence on world ecosystems, is an essential component of an international program.

Although the U.N.'s world-wide programs can provide the means for information exchange and global monitoring, mechanisms still need to be developed for: defining criteria for evaluation; establishing procedures; identifying the specific urban areas to be monitored; coordinating research and exchange of personnel and data; synthesizing and evaluating results; and developing policy recommendations. These complex functions may be too diverse to be handled by a single institution. Policy development, in particular, must be the concern of intergovernmental organizations. Creation of the other mechanisms we have mentioned could be conducted under the auspices of inter-governmental programs such as UNESCO's Man and Biosphere Program (MAB) or non-governmental science organizations such as the International Congress of Scientific Unions' Scientific Committee on Problems of the Environment (SCOPE).

The identification of common bases for comparative study of urban systems is central to the creation of an international monitoring program. In practical terms, action must be taken to:

1. Develop common methodologies and criteria for monitoring various aspects of the urban system, including: air and water quality, consumption of agricultural products, analysis of the effects of new technology, regional development programs, microclimatic variations, waste production, migration and settlement patterns, birth and death rates, population densities, presence and size of urban pest populations, incidence of disease, and other health indicators.

2. Identify specific urban areas in participating countries where such monitoring programs are possible, perhaps in conjunction with national or intergovernmental programs such as MAB or the U.N. Demonstration City Program.

3. Establish an operational procedure for the exchange of personnel, the synthesis of information, and the comparison of effects various urban policies have on components of the urban system.

The U.N. Conference on Human Settlements, scheduled for Vancouver, British Columbia, in 1976, offers an opportunity to develop such an exchange, particularly for the Demonstration City Program. The establishment of common criteria and methodologies for analyzing the cities selected for monitoring is critical to the implementation of the program.

Existing intergovernmental agencies, such as the Urban Sector Group of the Organization for Economic Cooperation and Development (OECD), the Organization of American States, the Center for Urban Information, or the Urban Research Institute of the Organization of Arab Cities, could jointly become an international mechanism for the exchange of necessary and relevant information. The Institute of Ecology (TIE), a non-governmental organization, should expand its Environmental Science Referral Service to include urban specialists as a specific category and broaden its Urban Ecosystems Program to include the entire western Hemisphere. Similar regional or multi-national programs should be encouraged elsewhere in the world.

Many existing bilateral programs can help provide links between nations and help assess the value of transferring environmental information from one cultural and economic climate to another. The join U.S.-Mexico Community Environmental Management Study (under the Center for Urban and Human Development in San Diego, California) and the Canada-U.S. International Joint Commission, which supervises international use of the Great Lakes, can provide guidance in developing programs. The recently signed agreement for scientific exchange between the U.S. and the U.S.S.R. designates certain urban areas in each country for the possible exchange of planning personnel and information. Within the U.S., a national program of urban system monitoring and assessment should be developed to contribute to and capitalize on the U.S.-U.S.S.R. exchange agreement and the U.N. Demonstration City Program.

In geographically diverse countries, data will need to be collected and analyzed for a number of separate regions; still, many policy issues can be effectively handled only on a national scale. A mechanism should be created to review and synthesize the data from regional analysis centers which deal with such inter-regional or national questions as: 1) immigration laws, public health programs, welfare, taxation, and other policies affecting population growth rates and distribution; 2) revenue sharing or other tax dollar allocation schemes which affect the economic vitality of urban areas; 3) location of major federal installations such as the Manhattan Project which stimulated the construction of Oak Ridge, Tennessee, or the NASA headquarters near Houston, Texas, both of which contributed to the size and rate of growth of those urban areas; 4) location and type of transit systems and other factors which affect urbanization patterns; 5) mortgage interest rates, real estate laws or federal policies affecting construction practices and urban form; and 6) the location of national parks and historical, cultural zoological botanical or recreational areas which may improve the quality of urban life by providing urban man visible links with his regional biota.

In smaller or more homogeneous countries, the policy considerations we have described might be combined with the collection and analysis functions on a regional scale (Level II).

Level II: State, Provincial or Multi-County Level

Because regional ecosystems may differ significantly from one another, regional natural resource and socio-economic analyses must differ to some extent in method and level of detail. These regional differences, however, are a matter of degree rather than of substance. For example, planners in all areas need the capacity to forecast economic activity and—given the requirements of the National Environmental Policy Act—to assess environmental impact. A common core of analysis and data requirements can be identified for different problems at the regional level, with local adaptations to meet unique regional conditions.

Comprehensive environmental analysis and data acquisition require the combined work of many specialists: ecologists, economists, political scientists and land use planners must all be involved. We need: economic analysis to predict economic effects of proposed actions; political analysis to describe the relationships between population change and management strategies; land use analysis to examine the allocation of regional space if a proposed action is taken; and ecological analysis to describe the ecological ramifications of proposed management alternatives. A common data base will facilitate such interdisciplinary study.

The structural nature of the regional institution which should undertake interdisciplinary research is not altogether clear. A semi-autonomous public agency, commission, board, or crown corporation seems more suitable than a large department or ministry. The scope of the agency given jurisdiction may be broadened beyond responsibility for environmental information to include other functions such as landscape reclamation and rehabilitation. The group might also engage in pilot demonstration projects to test new environmental markets, thereby stimulating business investments in such enterprises. The agency might promote the use of vacant urban land for activities such as maple syrup production, dog training, rabbit farming, or organic farming to allow urban residents to develop some "rural" skills. Whether or not the suggested agency should undertake activities beyond environmental analysis depends on local needs and perceptions and on the success of experimental pilot projects.

Level III: County or Metropolitan Level

At the county or major metropolitan scale, effort must be directed toward furnishing comprehensible environmental information to the local population. Sophisticated regional systems analysis techniques (Level II) are indispensable, but these should be supplemented by other methods, especially if the information is to be useful and accessible to local citizens. The "McHarg" technique is a useful public participation technique. It allows detailed environmental data to be presented in a relatively simple fashion to various audiences (McHarg, 1969; Dorney and George, 1970).

The McHarg methodology involves, first, a detailed inventory of an area's natural components and present and past land uses. The natural inventory can contain any number of items—surface and bedrock geology, topography, soils, hydrology, vegetation and wildlife, etc. Once the basic data has been collected, a value rating or interpretation of the natural resources is undertaken to clarify the spatial relationships of the elements identified. The data can then be used to assign potential use categories to various parcels of land: agricultural, forested, recreational, open space or urban. These ratings are determined by use of sets of prescribed criteria which serve to delineate the limitations of the land for certain potential uses. Studies of the visual qualities of the landscape have also been developed as a further means of assessing the land's suitability for particular uses. After the development potential of an area has been determined in this fashion, a political decision on land uses can be made. The political decision encompasses, of course, social, political and economic considerations as well as environmental realities.

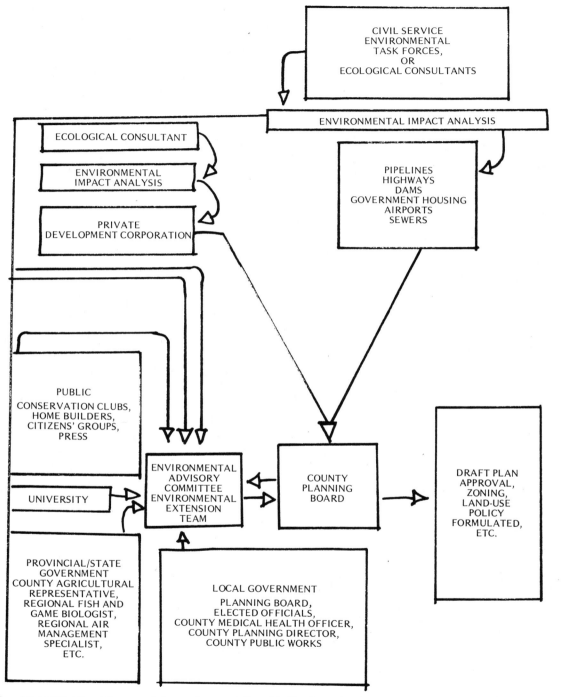

Figure 1   An organizational framework for environmentally guided land-use planning at the county level of government

Experiences in Minneapolis and Waterloo suggest that counties or metropolitan areas should retain at least a two-member environmental extension team to carry out county level planning. One member should have a thorough background in the physical and biological sciences; the other, a background in the design professions of architecture, landscape architecture or planning. Such a combination of expertise would be capable of utilizing and communicating ecological information to create and maintain healthy urban environments. The environmental extension team could respond to the needs of local residents and would be qualified to work with public and private groups at other levels, i.e. both the regional and the urban.

The team needs to be sufficiently independent from private and governmental pressures to establish credibility with private groups and individual citizens; at the same time, it must remain technically credible to the agencies and politicians with whom it must work. An example of such a regional planning scheme (Fig. 1) demonstrates how environmental information could flow through an institutional arrangement.

Level IV: City or Town Level

At the urban level, the range of information and analysis that is useful differs from that useful at other levels. Factors such as noise levels, parkways, recreational and cultural facilities and open space are often as important as environmental considerations—bird populations, plant species and diversity, water quality, air quality, etc. (Jacobs, 1970). Present urban planning practices tend to deal with natural resources and social concerns such as housing, public works or transportation as separate or semi-separate components. The conceptual analysis in the Waterloo, Ontario, study demonstrates, however, that they may not be separate. Excessive noise, poor water quality, unsightly buildings and neighborhoods, and lack of open space all occur in one place—the old downtown city core (Dorney, 1973). Although cause and effect relationships among these components are not clear, a decision to proceed with a proposed public work that would further deteriorate water quality and increase traffic through the downtown area would clearly be ill-considered. A city-level analysis which places social issues in an ecological framework would allow better judgments to be made than are presently possible.

McHarg has done such studies in Philadelphia, identifying coincidental patterns of natural resources and socially deprived areas in the city (McHarg, 1969). Such identification and speculation about the cause and effect relationships which give rise to observed patterns should be one dimension of the field of urban ecology.

An example of one coincidental pattern is the incidence of carbon monoxide levels in the downtown area and abnormal oxyhemoglobin and aspirin levels in blood taken from cab and truck drivers killed or injured in accidents. Dr. Cottrell, a medical health officer in Calgary, Alberta, has indicated that drivers may get headaches from poor air quality, perhaps aggravated by smoking or a faulty muffler; they take aspirin to relieve the headache, which further suppresses oxygen-carrying capacity in the blood. The result is impairment of physiological function which leads to higher accident rates.

To effectively use a city-wide system of environmental information, a small staff of environmental scientists and designers is needed (perhaps 2 staff members per 100,000 population). The staff would interpret information from the county or major metropolitan level (Level III) and explain it to municipal officials. The team would also conduct environmental analyses relevant to the needs of city residents. Environmental and social impact studies would also be circulated to this technical group for comment and would be kept on file in its office.

It is not clear from existing case studies precisely how such an environmental team would function in relation to other city departments. The team would have to have sufficient independence from planning and engineering staff pressures to voice opinions which may be contrary to others in city hall, those of developers, or of citizen pressure groups. They would need to have a fair degree of independence or buffering from local government pressures and be able to call on higher levels (Level III, II or I) for information requiring greater skill or more money to obtain than could be mustered on their own level.

The Dangers of Using Computer Information Systems

Most citizens do not understand computer procedures and assumptions and even many of those familiar with computers are unaware of hidden value judgments in computer programs. Computer information systems, accordingly, represent a potential infringement on citizen participation when they are used in making decisions which affect the public's daily lives and property. Special interest groups could influence the development of whole cities, town or regions and hence obtain an inordinate share of natural or economic resources by manipulating a computer system. Consequently, some inspection procedures and mechanisms of public control ought to be incorporated into computerized environmental information and analysis data systems before they are put into operation. Such controls might be similar to the audit procedures used to insure that private funds are not misused by state and federal bankers. Such audit could reduce both inadvertant and purposeful manipulation of the information system. Similarly, world environmental data systems should be similarly monitored by the United Nations to prevent the development of "information empires" which could hold at a disadvantage the smaller countries unable to staff and finance counter-systems.

Environmental Performance Standards

The benefit of using environmental performance criteria instead of land use zoning in urban planning projects clearly emerged in the Minneapolis study. The use of performance criteria allows more flexibility than the use of zoning since the criteria specify a minimum acceptable level of biological function or aesthetic quality for any particular unit of landscape. Planners can analyze changes within specified performance limits rather than in conformance with a rigid standard, as in the case of zoning. To formulate meaningful performance criteria, however, environmental data banks must be constructed for that purpose.

For example, performance criteria will define the opportunities for and constraints on achieving a goal such as retaining trout populations in a certain portion of a stream and suggest the costs associated with possible alternative actions. Performance criteria are distinguished from planning or design standards since they describe a series of probable cause and effect relationships rather than a limiting condition which ignores the relationships that have created that limit.

Performance criteria may fulfill two needs. First, there is the need to utilize the information network developed in planning and design processes. Performance criteria may allow the combining of data and data modeling on one hand, and the socio-political development process on the other. Second, performance criteria may meet the need to provide both incentives and controls that are anticipatory or offensive rather than reactive and based solely on past negative experience.

If the data gathering and analysis process at all levels of the information network are expressed in terms of flows, costs in energy dollars and social satisfaction, relative levels of stability, and relative levels of risk, then the limiting conditions and relationships between these conditions can be used to describe a range of action alternatives.

The larger the area or the larger the range of alternatives, the larger are the number of possible solutions to be investigated and the range of choices available. Performance criteria might also help to identify the best of the alternative actions.

The use of performance criteria may influence the socio-political decision-making process in the following ways: 1) by expressing opportunities and constraints in a multi-dimensional fashion, taking into account social, natural and psychological criteria; 2) by focusing on the relationships between dimensions rather than focusing on each as a separate and unrelated factor, as is traditional; 3) by formulating a range of choices for possible action rather than insisting on a single, best solution or reducing choice to a minimum: 4) by formulating opportunities and constraints based on what is now known without prejudicing the use of what might be learned in the future; and 5) by establishing common bonds between various environmental information networks to elucidate the limitations of proposed actions.

How to Use Environmental Information: A Political and Administrative Framework

Residents of some particular areas of a country may be more aware of issues related to environmental quality than residents of other areas. In communities where the level of public awareness is so low that environmental considerations have little or no effect on policy decisions, obtaining public money for planning studies or for the comprehensive analysis of an urban ecosystem will be difficult. In these cases, external stimuli are needed before the population will become aware of and evaluate the condition of their environment, and before they can begin to visualize better conditions for the future.

To achieve this initial stimulation, the American Institute of Architects has developed and refined a program called RUDAT, Regional Urban Design Assistance Team (Bara, 1970). The program, developed over a six-year period, provides for a team of experts to briefly visit a community to advise in planning. Many organizations have used this approach for widely different purposes and with varying degrees of success. The U.N. Development Program, for example, uses teams of experts for many assignments. The city of Halifax, Nova Scotia, used a team of encounter group specialists to stir community interest in the environment (Metropolitan Area Planning Commission, "Encounter Reports"). The methodology is simple and inexpensive. Although the approach can be misused with detrimental results, it can be an effective tool for purposes of stimulating community awareness and for catalyzing action if certain requirements are met; the composition of an assistance team will vary, but the following guidelines may be helpful in gathering a team: 1) Individuals with training and practical experience in more than one discipline are often effective members of such a team. 2) For international projects, individuals who have worked for an extended period outside their own culture are preferable to well known experts. The latter usually have reputations established in a single discipline and international experience only in a consultant capacity. 3) The team should always include natural and social scientists as well as those actively working on urban problems. 4) Individuals with training in fields such as geography and cultural anthropology may help bridge communication gaps between the social and physical scientists. Architects and landscape architects often have the general training needed to provide the bridge between scientists and such practitioners as engineers and planners; economists or lawyers appear best suited to facilitate communication between scientists and politicians. 5) The team should be capable of visualizing and portraying alternative cultural and physical conditions to the community.

The logical evolution of the RUDAT concept, suggested by experiences in Minneapolis, Halifax, and Waterloo, may be that at some point a small, permanent nuclei of professionals from the environmental sciences, social sciences, and design field should be hired to assist the city, county or regional government. The group, called an "environmental extension team", may best be associated with an Environmental Advisory Committee (Fig. 1). Experience in Waterloo County suggests that the team could work effectively with various city planning and city engineering departments at the county or regional level of government. The team could also have close ties with regional, state or provincial, and federal resource management agencies such as the Geological Survey, Weather Service, or Forest Service in the United States, or the ministries and departments of Agriculture, Public Works and Environment and Natural Resources in Canada. Civil servants of these agencies, residing in particular locales, should be appointed to the Environmental Advisory Committee; their appointment would facilitate the transfer of information among these agencies and between them and public pressure groups and the media, etc. The Environmental Advisory Committee members could address themselves to controversial environmental issues and their recommendations could flow to city councils, regional councils, or higher levels of government, depending upon the political and financial climate.

In Waterloo County, Ontario, we have completed four years of experience with an ad hoc environmental advisory committee reporting to the planning board (Planning and Development Committee). Our experience demonstrates that academic environmental quality mapping can be transformed into a realistic draft of an official plan in about 12–18 months. The experience has also shown that the team can address contentious environmental issues (in this case, controversy over a major dam) and can demonstrate by joint study resolution the need for more specific environmental (hydrological) informa-

Such an advisory committee could be socially responsive since it provides a democratic forum for debate, discussion, and resolution of conflict. It gives public and private interest groups a forum before a diverse committee capable of rebutting unfounded, distorted or false statements. The committee process, which is political, identifies the issues, the interest groups, and the trade-offs that must be made before economic and political conditions require rapid and irreversible action.

The committee process is economically sound since it may avoid losses in time and money resulting from injunctions and protests; it can avoid wasteful expenditure of development money resulting from inadequate or inappropriate knowledge. In Waterloo County, for example, millions of dollars worth of funding have been proposed for construction and development; such projects often have undesirable primary and secondary environmental effects. With adequate environmental impact analysis, and with committee evaluation through the mechanism suggested here, management of urban and regional land and of human and economic resources can be improved considerably (Dorney, 1972; Coleman and MacNaughton, 1971).

THE BUTTE, MONTANA, RUDAT STUDY

Background on RUDAT

The objective of a three-day RUDAT reconnaissance visit to Butte, Montana was "a plan for planning". The visit was made in June of 1972 by a team of nationally recognized scientists, planners and designers assembled under the RUDAT program (Regional/Urban Designers Assistance Team).

The American Institute of Architects' RUDAT program was initiated in June, 1967, with the visit of a small team of architects to Rapid City, South Dakota. Since then, 13 other communities from Davenport, Iowa, to Akron, Ohio, and Fairfax County, Virginia, have had RUDAT teams visit and advise them. The methodology of the program has evolved gradually and the composition of the teams has been broadened considerably since 1967. In Butte, the RUDAT team was headed by Milwaukee architect and planner Maynard Meyer and included two other planner-architects, two urban designers, a regional plant ecologist, and a geologist.

The RUDAT approach operates on the assumption that persons not familiar with longstanding problems can see a community through new eyes and can find fresh approaches to stimulate community awareness and to instill new enthusiasm in local professionals who have been bogged down by the day-to-day details. The RUDAT team visits an area for only three days and looks only at major issues. It cannot provide a detailed analysis of problems, nor detailed solutions to them.

However, the team can be innovative, opening new possibilities for future improvements and providing the basis for future in-depth studies on which the community and interested governmental agencies can focus attention. The team's visit may create enthusiasm among residents as a stimulus to action.

In preparation for a RUDAT visit, the community representative, usually the local planning board, assembles any data, reports or other documents to give the team concise background information about the community. These items usually include: census data; U.S. Department of Agriculture Soil Conservation Service maps; U.S. Geological Survey topographical maps; surface and sub-surface ecological data; low-level aerial photos of the area; vegetation maps; ground water surveys; local weather data; existing development plans; and previous studies of the area.

After reviewing the background material, the team surveys the area first-hand and meets with local officials, industrial, business and union leaders, and representatives of local citizens' groups. Intensive discussion with these groups assist the team's assessment of the community's critical issues and its formulation of recommendations for community action, or its "plan for planning" (see Fig. 2).

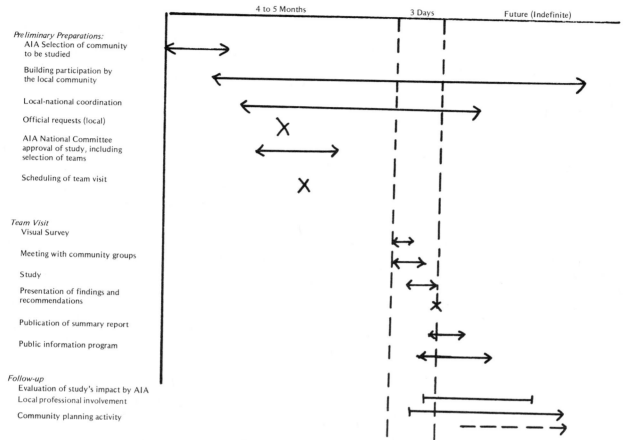

Figure 2 The time frame of the Rudat process*

*"Checklist for Cities" Doc. U102, AIA, Washington, D.C.

The team stimulates community action by: 1) contacting all major community agencies and public groups prior to the visit; 2) meeting with a wide cross-section of the community once it has arrived; 3) having local radio, television and newspaper coverage of its visit; 4) publicly presenting its recommendations; 5) participating in an in-depth television interview about its findings; 6) publishing its recommendations; and 7) contributing to a series of in-depth news articles on those issues highlighted in the team's report.

The Problems of Butte

Butte, Montana, is a mining community situated in Silver Bow County west of the main range of the Rocky Mountains known as the

Continental Divide. The city has two significant geographical characteristics: the older section of the city is built on the side of a mountain; the newer developing area is located on the "flats" to the south.

The site of the "Richest Hill on Earth", Butte has known great wealth, fantastic promise, conflict, spasmodic growth, and intermittent decline and prosperity. Early growth in Butte was characterized by lack of city planning or the development of amenities. Extreme weather and working conditions encouraged most settlers to locate around the mine shafts on the hillside; such development has left the city with a collection of older, closely built homes constructed prior to 1929. Approximately 65 per cent of these homes are now estimated to be sub-standard.

Butte is still a mining city, and its prosperity fluctuates with the industry. Unionism, another major socio-economic influence in the area, is strong; nearly fifty unions exist in Butte.

The future of Butte will be significantly influenced by two additional factors: the Berkeley Pit; and the interstate highway system. Butte's location at a north-south east-west intersection of the interstate highway system gives the city the opportunity to develop new jobs and businesses, especially for services and recreation. The Montana College of Mineral Science and Technology, one of the nation's finest technical colleges, is also located in Butte.

In 1864, two prospectors from nearby Virginia City, Montana, found placer gold along the banks of Silver Bow Creek; mining claims quickly opened and shafts were sunk. The miners built their shacks, cabins and modest houses near the diggings they worked. Hundreds of these homes, still occupied, dot the hill and dip locations near the gallow frames of working mines.

Mineshafts cross and crisscross beneath the topsy turvey streets of Butte. If all these mine shafts and tunnels were placed in a single line, they would traverse the United States.

A labor dispute idled 5,700 copper industry employees in the Butte area from June 1967 to March 1968. Silver Bow County and many parts of Montana have been hit periodically by such strikes.

Butte's metropolitan area has a population of 42,000. A few generations ago, the population was 80,000 to 90,000. During the World War One period, 17,000 men worked in mines in the Butte area; now, through automation and the introduction of mechanized open pit mining, the underground mining force numbers 1,500. There is, however, no feeling of despair in the community. Rather, there is keen awareness of the need for community-wide cooperation in planning and action to develop the economy and for educational, health, public welfare, transportation, recreational, and housing facilities, and improved city-county relations and programs. One solid community goal is the elimination of blight.

Mining perhaps has had its greatest effect on the life of the community through the operation of the Berkeley Pit, opened in 1955. The Pit is an enormous, ever-widening cavity, now more than a mile across measured east to west. Now nearly 1,400 feet deep, the Pit will eventually reach 3,000 feet or more. It is located only a few blocks from the business district of Butte.

Mining continues to be the major industry in Butte. In 1966, the Anaconda Company handled 58.6 million tons of rock to produce 17.3 million tons of ore from which 240 million pounds of copper emerged. The city shares the company's concern about whether such mining operations can continue to be profitable. If jobs and business are to continue in Butte, investment in mining must continue to return a profit.

Butte residents feel that their future also depends on development of trade. The city's location as a hub on the transcontinental interstate highway system favors such development.

The major city in western Montana, Butte is surrounded by mining, farming, timber and livestock operations. Butte provides well-equipped and well-staffed hospital and medical clinics and legal and educational service throughout its trade area. News and entertainment are furnished by the city's newspapers, radio and television stations.

Results of the RUDAT Study in Butte

As a result of the RUDAT study in Butte, Montana, several actions have taken place. The city of Butte and the Anaconda Copper Company have begun to discuss a new building complex; for several years, there had been no discussion on that issue. Anaconda will build a new company office and will provide funding to relocate municipal facilities in a jointly-selected site.

Planning consultants have been commissioned to study Butte's central business district; the business area will be retained and refurbished and will include a buffer zone of open space between the mining pit and the shopping area.

The city and the regional government have begun to discuss the negative effects of scattered urban development and water pollution and are considering possible solutions.

Such developments, within a few months of the study, are positive and encouraging.

There is a danger in the RUDAT "environmental animation" approach. Advice offered on the basis of short-term study could result in damaging social or environmental changes. Protection against such a possibility depends on adequate monitoring of past RUDAT activities by the AIA or other involved organizations. The inclusion of geologists and ecologists on the American Institute of Architects' RUDAT team is commendable. The balance of future teams could be further improved by the addition of sociologists, urban anthropologists, and landscape architects.

THE MINNEAPOLIS-ST. PAUL STUDY

The Minneapolis study, conducted by the consulting firm of Wallace, McHarg, Roberts, and Todd, used an ecological, design, and planning approach (Wallace, McHarg, et al., 1969). The study was commissioned by the Metropolitan Council of the Twin Cities to deal with the problems resulting from decades of urban and industrial growth which had little relation to the natural resources of the area in which growth was occurring. The study, completed in June of 1969, evaluated the Minneapolis-St. Paul metropolitan region in terms of a synthesis of its physical natural resources—its physiography, geology, climate, hydrology, and plant and animal life. These resource analyses became part of an overall planning approach leading to recommendations on specific sites (see Fig. 3).

The study was one of the first of its kind. In the introduction to the report, McHarg writes, "The planning process asks nature to advise man as to the form and pattern of metropolitan growth... The purpose of the ecological plan is to inventory phenomena and natural processes, to reconstitute these as a value system, and to perceive the degree to which land, air and water processes offer both opportunities and restraints to single and combined prospective land uses...." (Wallace, McHarg, et al., 1969).

The Minneapolis study is a guide rather than a plan. The layout of streets, gutters and

196 Part II

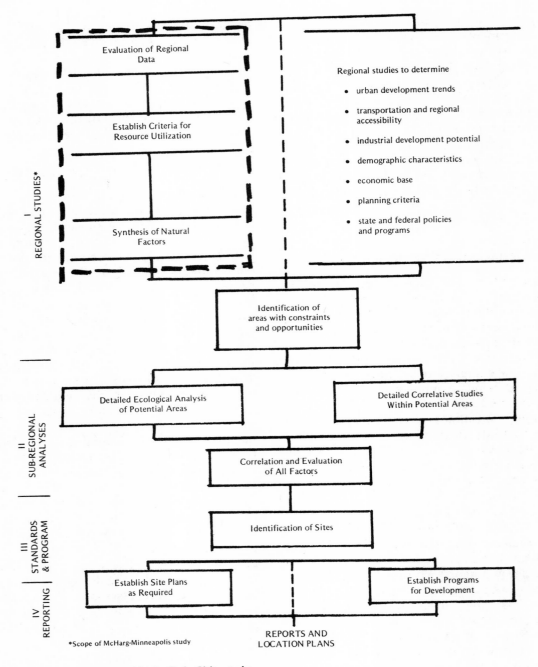

Figure 3 Scheme used in the Twin Cities study

sewers is clearly left to city planners. The study was intended to speak for the land, to let decision-makers know where the city can and cannot go.

In the three years since the study's completion, five steps have been taken. These illustrate the effectiveness of the study.

1) An open space protection act is presently in preparation.

2) The flood plain sanitary land fill, which had been causing water pollution, has been closed.

3) Solid waste disposal activities will be moved to sites more suitable in terms of the geology and soils of the area.

4) The issue of alternative airport locations was examined and a second study was commissioned.

5) Environmental performance standards related to environmental capability and suitability of the area, outlined in the study, have been put into use. These standards have allowed the Metropolitan Council to veto any inappropriate development proposed by the separate city councils. Developers are now following the environmental performance standards to avoid veto by the Metropolitan Council; as a result, developers are becoming more environmentally aware and are hiring competent professionals to guide their decisions.

THE HALIFAX, NOVA SCOTIA, STUDY

The study of Halifax, Nova Scotia, was generated by concern over the consequences of relocating the major water supply coming from two metropolitan-owned watersheds immediately adjacent to the urban center. This availability of 4,179 acres offered an unusual opportunity to explore planning alternatives for urban development (Jacobs, 1970); the land was owned entirely by the city, avoiding difficulties with privately-owned parcels.

A planning process (see Fig. 4) exploring alternative uses for the vacated site considered social and economic development demands. A "development framework" was used to examine the site's resources and land use possibilities. A network of open space through the study area was used as a gaming board, allowing a variety of community groups to explore their own development objectives, strategies and location priorities on the site.

The areas of conflict and agreement among the groups were recorded and were used to develop a program resulting in an alternative design.

Since completion of the study, a metropolitan-wide planning project has been started, using the open space framework developed in the study, but expanding and modifying it to include planning for the entire metropolitan region (Metropolitan Area Planning Commission, 1973). Public decision-makers at the municipal and provincial level and area citizens have been given the opportunity to examine development options on the publicly-owned watersheds. Such examination is critical before irreversible actions are taken to sell the land or to fund expensive water, sewer or power services to the area without regard for the effects on urban form or urban quality.

THE WATERLOO, ONTARIO, STUDY

The study of Waterloo, Ontario, arose from a desire to develop an inexpensive, quick and understandable analysis of urban natural and amenity resources (see Fig. 5). Members of the faculty of Environmental Studies at the University of Waterloo believed that if such an approach were successful, it would stimulate similar studies elsewhere. They also felt the approach might instigate a process where-

198 Part II

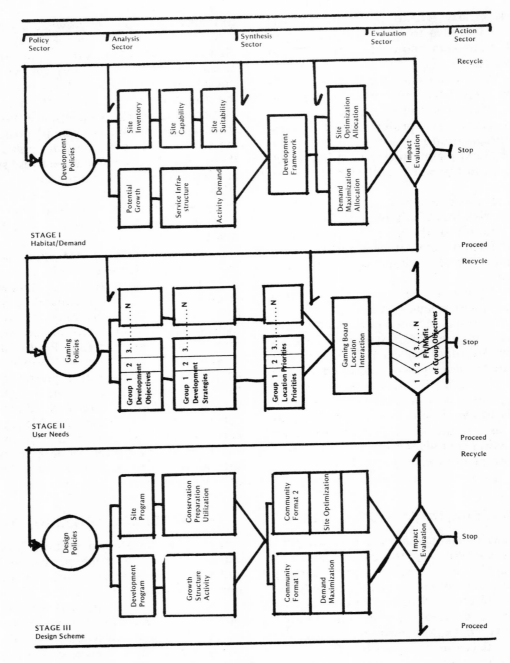

Figure 4   Scheme used in the Halifax case study

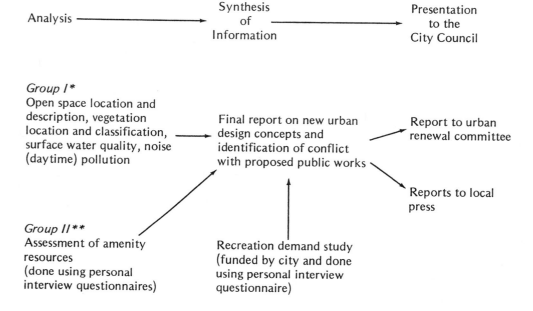

*Group I — graduate students in urban and regional planning
**Group II — third-year undergraduate students in geography

Figure 5
Method used to study issues of urban environmental quality in Waterloo, Ontario

by information on urban environmental quality was regularly gathered and made available to city councils facing decisions on related matters.

Waterloo is a small town with an agricultural and light industry orientation. The two universities in the town have a combined enrollment of 15,000. New suburban areas close to the universities contrast sharply with the older residential sectors of the city.

The study of Waterloo began in January, 1972, and was completed by two groups of university students in April, 1972. The questionnaires used in the study were designed so they could be updated at low cost ($5,000) to the city. The product of the study was an interim report to the Urban Renewal Committee of the City Council and the City planner, Donald Scott, who cooperated with the students working on the project.

The City Council has prevented the widening of one tree-lined street in the downtown area; community groups in and beyond the specific area involved, using information gathered in the study, became aroused over the environmental ramifications of the proposed project and took action to make their viewpoint heard in city hall.

In addition, since the study a breeding bird count has been completed; it provides an overview of avian diversity in relation to man-made and natural habitat types identified in a vegetation survey.

The completion of the first report led the city to allocate $5,000 for further work on urban space in relation to recreational demands.

Other recommendations in the Waterloo study are apparently in conflict with proposals made to the city by a major developer and by the Grand River Conservation Authority, which has recommended that a long box culvert be built in the downtown section of Laurel Creek, a small stream looping through the city. Whether or not these conflicts can be resolved remains to be seen (Dorney, 1973). However, the potential adverse effects of the proposed projects have been identified before decisions have been made, opening the way for dialogue between citizens' groups and the City Planning Board and City Council.

THE KNOXVILLE, TENNESSEE, STUDY

The Tennessee Valley Authority region seems a suitable area to conduct regional planning studies since there is extensive socio-economic data going back many years, as well as scientific studies of the area's natural environmental capabilities made by the Oak Ridge National Laboratory.

Agencies in the Tennessee-Kentucky area have been engaged in regional analysis. The Tennessee Valley Authority, regional commissions, state agencies, and in recent years, local development districts have been actively making economic, demographic and, in some cases, ecological forecasts. With the advent of the National Environmental Policy Act of 1969, court interpretations of the act, and the 1972 amendments to the Federal Water Pollution Control Act, agencies not previously required to consider the environmental effects of their actions must now do so. Although federal, regional, state and local agencies are required by the National Environmental Policy Act to prepare comprehensive impact appraisals, they often lack any comprehensive impact appraisal methodology.

In June, 1970, the Regional Environmental Systems Analysis (RESA) program was initiated as part of a National Science Foundation program on Environment and Technology Assessment at the Oak Ridge National Laboratory in Tennessee. The RESA program has been used to develop analytical tools to help government agencies charged with planning and management responsibilities address substantive questions. The use of RESA by such agencies has identified methodological insufficiencies and provides opportunities to validate and refine the analysis.

The Tennessee Valley Authority, the Tennessee State Planning Office, the East Tennessee Development District, and the Appalachian Regional Commission are the primary agencies whose planning capabilities may be complemented by the RESA program's research. Working arrangements between the RESA program and the technical staffs of TVA and the East Tennessee Development District have already been established.

The East Tennessee Development District, first chosen for study, covers 6,500 square miles, and includes 16 counties with a total population of approximately 700,000 people. The Knoxville Standard Metropolitan Statistical Area lies within the study area.

Under the RESA program a conceptual plan was developed for analyzing the region's environmental systems. The major research concerns were defined as the socio-economic, land-use, ecological and socio-political areas (See Fig. 6). The relationships of these areas

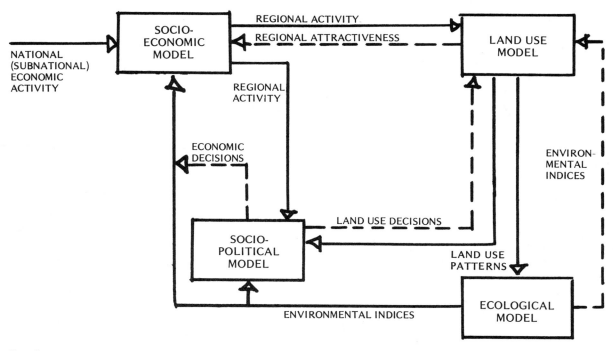

Figure 6
Regional environmental systems analysis (RESA) model developed for Knoxville, Tennessee

(Olson, C. Unpublished data. Oak Ridge National Laboratory)

of concern were examined and the disciplinary skills and data required for the study were collected.

Socio-economic Analysis. A macro-simulation model describing the regional socio-economic system was developed to assist in making regional population and employment forecasts. Trend projection, cohort-survival analysis, and export-base analysis techniques were combined in the model which incorporated shift-share analysis as well. The study approach was empirical, using weighted regression analysis of pooled cross-section data (Olson, 1972). The present model was calibrated by taking a known population and employment base (1960 census data) and using the model to make yearly projections of the population (by age, sex and social group) and the employment (by industry) to another known data base (1970 census data).

Land Use Analysis. Land use forecasts for aggregate regional activities were made by developing a spatially deterministic model which allocated primary and secondary employment first to homogeneous sub-regions and then to 170-acre cells in the East Tennessee Development District. The process was tested using a detailed flow chart and hypothetical data.

Land use is assigned on the basis of the "attractiveness" scores computed for each sub-region and cell. The scores are calculated by developing indices from both subjective and statistical data using expert advice and factor and cluster analyses. The indices are weighted according to a Delphi analysis—i.e., objectives are rated by knowledgable groups. Because of the information handling needs of the land-use model, the Oak Ridge regional modeling information system, a regional analysis data system, has been developed. The necessary land use data have been collected, classified, digitized and stored in the system.

Ecological Analysis. To describe the transport, deposition, and effects of effluents in the environment, existing air and water transport models for the region were modified and aquatic and terrestrial ecosystem models were developed in cooperation with other agencies. An air transport model was modified to describe SO_2 and particulate distribution as a function of their sources and atmospheric characteristics. Two water transport models were adapted to simulate changes in the flow and water quality characteristics of streams and rivers caused by land-use changes in the watershed. A terrestrial ecosystem model was developed and considerable data collected; an aquatic ecosystem model was also developed, to simulate the impact of flow rates, water quality and land use on the aquatic plant and animal population (Williams, 1971; Leckie and Ulrich).

Socio-political Analysis. A public budget model and a political interaction simulation program were acquired and developed to depict the relationship between residents and the regional governing system. The public budget model will simulate the supply of and demand for public goods and services in each of the major political jurisdictions of the East Tennessee Development District; it will ultimately be able to forecast major economic changes. The political interaction simulation program, on the other hand, will forecast probable political responses to proposed economic, land use, and ecological policy changes (Leckie and Ulrich; Bulkley, 1971).

THE NORMAN, OKLAHOMA, MODEL

The urban-rural ecosystem model developed in the Norman, Oklahoma study is conceptual only and not operational (Table 1). The study does serve, however, as an example of a theoretical approach somewhat similar to the RESA study in Knoxville (Parton, 1972); it combines urban resource uses, rural resource production linked through regional climate on a seasonal basis, and other rural land uses. The land component would be controlled by the ecological restraints built into an operational social system having computer capability.

SUMMARY

Our review of these case studies, and our past experience, suggest that there is presently enough theoretical and factual information and enough professional expertise available to begin limited but effective management of urban environments in a way compatible with ecological principles; such management can begin now. Better urban ecosystem management depends on adequate qualitative and quantitative mapping of natural resources, on local environmental extension teams, on citizen environmental advisory groups, and on the replacement of zoning standards by environmental performance criteria.

Minneapolis, Oak Ridge, Louisville, Halifax, Waterloo (Ontario) and Vancouver are a few of the places in North America with adequate personnel and environmental data to serve as pilot areas for the implementation of ecologically sound urban management policies.

Environmental information systems should include adequate public safeguards. Decisions affecting tax bases, life styles, land use and the historical and cultural continuity of urban areas should not be altered to serve illegal or selfish ends, nor to conform to the predetermined goals of technocrats, however well-meaning. Urban environments would be better planned in the present ad hoc fashion than by management of our destinities by computer models whose judgments could be purposely manipulated by individuals or groups working for special interest goals.

RECOMMENDATIONS

Recommendations for Action

IV-1. We recommend the establishment of an environmental information system including the national, regional, county, and city levels. For initial development and assessment, a system based in the lower three levels (regional, county and city) could be established using existing expertise.

IV-2. We recommend that both urban and developing areas employ environmental extension personnel. Such personnel should form interdisciplinary teams supported at higher governmental levels by similar teams with computer and systems analysis capabilities.

Research Recommendation

IV-3. We recommend research on means for preventing the misuse of environmental information on the national, regional, county and city levels.

REFERENCES

American Institute of Architects. 1972. A checklist for cities. Doc. U102. Washington, D.C.

Bara, Vilma. 1970. Vehicle for urban improvement. American Institute of Architects Journal (Aug. 1970) 49–54.

Bulkley, Johnathan W. 1971. Modeling decision effectiveness: application to the water resource systems. Paper presented at the International Symposium on Mathematical Modeling Techniques in Water Resource Systems, Ohawa.

Coleman, D. and S. MacNaughton. 1971. Environmental planning in Waterloo County. Waterloo County Area Selected Geographical Essays, Univ. Waterloo, Geography Department Waterloo, Ontario.

Dorney, R. S. 1972. Ecological land-use planning in Canada: issues, methodology, public, professional and business interaction, and a strategy for implementation. Seminar Canadian Council Resource and Environmental Ministers, Fredericton, N.B. Sept. 19–22.

Dorney, R. S. 1973. Inventory, analysis and assessments of urban environmental quality: a case study of Waterloo, Ontario. Faculty of Environmental Studies: University of Waterloo, Waterloo, Ontario.

Dorney, R. S., and M. G. George. 1970. An ecological analysis of the Waterloo-South Wellington Region. Environmental Studies, University of Waterloo, Waterloo, Ontario.

Jacobs, Peter. 1970. Site planning process: activity allocation and design scheme. Nova Scotia Technical College, Halifax, Nova Scotia.

Leckie, Peter D., and Martin A. Ulrich. An operational multi-jurisdictional model of Vancouver's local public sector. First Year Report of the Local Government Group of the Inter-Institutional Policy Simulator Study, Resource Science Centre, University of British Columbia, Vancouver, B. C., Canada.

McHarg, I. 1969. Design with nature. Natural History Press, Garden City, New Jersey.

Metropolitan Area Planning Commission. "Encounter Reports". Department of Municipal Affairs, Halifax, Nova Scotia.

Metropolitan Area Planning Commission. Plan and

strategy for growth. Department of Municipal Affairs, Nova Scotia (in press 1973).

Meyer, Maynard. 1973. Butte, Montana RUDAT Report. American Institute of Architects, Washington, D.C.

Olson, R. J. 1972. Migration to and from state economic areas in the interior Southeast: an experiment using pooled cross-section data. ORNL-NSF-EP-17.

Parton, W. J., Jr. 1972. Development of an urban-rural ecosystem model. Meterology Department, University of Oklahoma.

Wallace, McHarg, Roberts and Todd. 1969. An ecological study of the Twin Cities metropolitan area. Philadelphia, Pa.

Williams, R. B. 1971. Computer simulation of energy flow in Cedar Bog Lake, Minnesota, based on the classical studies of Lindemann. Page 543 in B.C. Patten, ed. Systems analysis and simulation in ecology. Vol. 1. Academic Press, New York.

Institutional Affiliations

Advisory Council and Urban Ecosystems Workshop Participants 1973

Robert Anderson, Jr.
Department of Economics
University of California
Riverside, California

Lloyd Axworthy
Institute for Urban Studies
University of Winnipeg
Winnipeg, Manitoba Canada

Alan M. Beck
Center for the Biology of Natural Systems
Washington University
St. Louis, Missouri

Alexander Bigler
Alexander B. Bigler Associates
Oakton, Virginia

Sinclair Black
School of Architecture & Planning
University of Texas
Austin, Texas

L.J. Bledsoe
Center for Quantitative Science
University of Washington
Seattle, Washington

Institutional Affiliations

Arthur Boughey
Department of Population & Environmental
  Biology
University of California
Irvine, California

Richard Bradley
Department of Economics
University of California
Riverside, California

Kent W. Bridges
Department of Botany
University of Hawaii
Honolulu, Hawaii

William R. Burch, Jr.
School of Forestry
Yale University
New Haven, Connecticut

Hugh Calkins
Department of Civil Engineering
University of Washington
Seattle, Washington

Tim Campbell
Department of Urban Studies and Planning
Massachusetts Institute of Technology
Cambridge, Massachusetts

Daniel Carson
School of Architecture
University of Wisconsin
Milwaukee, Wisconsin

Alan D. Chambers
Resource Science Center
University of British Columbia
Vancouver, B.C. Canada

Adolf Ciborowski
Research & Development

Center for Housing, Building & Planning
United Nations
New York, New York

Robert T. Coles, AIA
Buffalo, New York

Ralph W. Conant
Institute for Urban Studies
University of Houston
Houston, Texas

Donald Conway
American Institute of Architects
Washington, D.C.

William Cooper
Department of Engineering Systems
Michigan State University
East Lansing, Michigan

Aryeh Cooperstock
Ministry of State for Urban Affairs
Ottawa
Ontario, Canada

George Cox
Department of Biology
California State University
San Diego, California

Clyde W. Craven
Regional Environmental Systems Alanysis
  Program
Oak Ridge National Laboratory
Oak Ridge, Tennessee

Ralph d'Arge
Department of Economics
University of California
Riverside, California

Craig Davis

School of Community & Regional Planning
University of British Columbia
Vancouver, B.C. Canada

David E. Davis
Department of Zoology
North Carolina State University
Raleigh, North Carolina

Edward Deevey, Jr.
Florida State Museum
University of Florida
Gainesville, Florida

Josephine Doherty
RANN
National Science Foundation
Washington, D.C.

Robert Dorney
School of Urban and Regional Planning
University of Waterloo
Waterloo, Ontario Canada

Joan Draper
Student Project Population Institute
Washington, D.C.

Richard Duncan
Center for Quantitative Science
University of Washington
Seattle, Washington

Amos Eddy
Department of Meteorology
University of Oklahoma
Norman, Oklahoma

Robert Einsweiler, AIP
Planning Consultants
Minneapolis, Minnesota

David Erickson
Division of Systems Research
Illinois Department of Conservation
Springfield, Illinois

Charles Ervin
University of Texas
Austin, Texas

William Garrison
Environmental Systems Engineering Program
University of Pittsburgh
Pittsburgh, Pennsylvania

James E. Hackett
Division of Environmental & Urban Systems
Virginia Polytechnic Institute and State University
Blacksburg, Virginia

Linda Harris
Department of City & Regional Planning
University of Pennsylvania
Philadelphia, Pennsylvania

William Hart
Department of Biology
Dalhousie University
Halifax, Nova Scotia Canada

Kingsley E. Haynes
LBJ School of Public Affairs
University of Texas
Austin, Texas

Lawrence Heidemen
School of Public Health
University of Texas
Houston, Texas

Patrick Horsbrugh
Environic Design Program
Department of Architecture
University of Notre Dame
Notre Dame, Indiana

Institutional Affiliations

Edgar Horwood
Urban Data Center
Department of Civil Engineering
University of Washington
Seattle, Washington

Peter House
Office of Research & Monitoring
Environmental Protection Agency
Washington, D.C.

Malcolm K. Hughes
Department of Biology
Liverpool Polytechnic
Liverpool, England

John Hultquist
Institute of Urban & Regional Research
University of Iowa
Iowa City, Iowa

Norman Hummon
Environmental Systems Engineering Program
University of Pittsburgh
Pittsburgh, Pennsylvania

Conrad Istock
Department of Biology
University of Rochester
Rochester, New York

Peter Jacobs
Faculte de l'Amenagement
Department of Architecture
University of Montreal
Montreal, Quebec Canada

David Jameson
Graduate School
University of Houston
Houston, Texas

Warren A. Johnson
Department of Geography
San Diego State University
San Diego, California

Royce LaNier
Director, INTECOL
University of Wisconsin
Madison, Wisconsin

Richard Lee
Department of Health, Education & Welfare
Bureau of Community Environmental
   Management
Rockville, Maryland

Charles Lewis
The Morton Arboretum
Lisle, Illinois

Ingrid Lundholm
The National Board for Urban Planning
Bromma, Sweden

Thomas MacCalla
Center for Urban & Human Development
United States International University
San Diego, California

Bassett Maguire, Jr.
Department of Zoology
University of Texas
Austin, Texas

Brian Mar
Water & Air Resources Division
Department of Civil Engineering
University of Washington
Seattle, Washington

Robert Mather
Department of Architecture
University of Texas
Austin, Texas

William R. McGrath
Raymond, Parish and Pine, Inc.
Transportation Planning Group
White Plains, New York

M.B. McPherson
ASCE Urban Water Resources Research
  Program
Marblehead, Massachusetts

Stuart Millendorf
Office of Environmental Programs
Port of New York Authority
New York, New York

Robert Murray
Houston, Texas

Reynell Parkins
School of Architecture
University of Texas
Austin, Texas

Clifford H. Patrick
Oak Ridge National Laboratory
Urban Research Section
Oak Ridge, Tennessee

Daniel Phelps
Department of Civil Engineering
University of British Columbia
Vancouver, B.C. Canada

James B. Pick
Department of Populations and Environ-
  mental Biology
University of California
Irvine, California

Alan Rabinowitz
Department of Urban Planning
University of Washington
Seattle, Washington

Martin Redding
Office of Research & Monitoring
Environmental Protection Agency
Washington, D.C.

William Rees
Community & Regional Planning
University of British Columbia
Vancouver, B.C. Canada

Stephen Reeve
Urban Ecosystems Project
University of Wisconsin
Milwaukee, Wisconsin

Theodore Roefs
Hydrology and Water Resources
University of Arizona
Tucson, Arizona

J. Barkley Rosser, Jr.
Institute for Environmental Studies
University of Wisconsin
Madison, Wisconsin

Paul Sedway, AIP
Sedway/Cooke
San Francisco, California

Chandler Smith
Duke Environmental Center
Duke University
Durham, North Carolina

C. Ray Smith
School of Urban and Regional Planning
University of Waterloo
Waterloo, Ontario Canada

Forest Stearns
Department of Botany
University of Wisconsin
Milwaukee, Wisconsin

Chandler H. Stevens
School of Architecture
Rensselaer Polytechnic Institute
Troy, New York

W. Cecil Steward
College of Architecture
Texas A & M University
College Station, Texas

Robert Stuart
Urban & Regional Planning
Virginia Polytechnic Institute and State University
Blacksburg, Virginia

Erik A. Svenson
Institute for Applied Technology
National Bureau of Standards
Washington, D.C.

Linda Alvin Thomas
Department of Botany
University of Wisconsin
Milwaukee, Wisconsin

Henry Vaux
Graduate School of Business Administration
University of California
Riverside, California

James A. Veltman
The Woodlands Development Corporation
Houston, Texas

D.H. Waller
Department of Civil Engineering
Nova Scotia Technical College
Halifax, Nova Scotia Canada

Robert Warren
Center for Urban Affairs
University of Southern California
Los Angeles, California

Anthony Wilden
Communications, Muir College
University of California - San Diego
La Jolla, California

Robert Williams
Office for Environmental Development
Soil Conservation Service
Washington, D.C.

Harold F. Wise, AIP
Planning Consultant
Washington, D.C.

Robert A. Woerner, ASLA
Spokane, Washington

Norman Woldow
Austin, Texas

James Wuenscher
School of Forestry
Duke University
Durham, North Carolina

Arthur J. Zeizel
Office of the Assistant Secretary for Research and Technology
Department of Housing and Urban Development
Washington, D.C.

James Zucchetto
Department of Environmental Engineering
University of Florida
Gainesville, Florida

Material prepared by the task groups of the Urban Ecosystem Workshop has been rearranged for use in this volume. The reports of the various groups appear here as follows:
Goals for the Urban Ecosystem: pp. 37–44; pp. 59–62; pp. 70–72; p. 78.
Institutional Structures: pp. 44–59; pp. 116–117.

Physical Structure and Function: pp. 85–98; pp. 117–120.

Resources: Water—Energy—Materials: pp. 98–107; pp. 120–138.

Population Processes: pp. 68–70; pp. 72–78; pp. 79–85.

Indicators of Urban Ecosystem Function and Health: pp. 138–163.

Case Studies of Cities and Urbanized Regions: pp. 180–204.

Systems Capacities, Limits and Intersystem Linkages: pp. 163–173.

Index

Accounting system, social-environmental, 59–60, 153–156
Advertising, 131, 132–133
Alinsky-type community organizations, 57
Allergies, plant-induced, 83–84
Alliances, 45
Animal bite, 83
Animal parasites, 83
Animals
  diseases, 83
  economic benefits, 84–85
  indicators of environmental quality, 84, 163
  psychological benefits, 10, 21, 82, 84–85, 110
  urban, 81–85, 110
Assimilation, biological, 150–151 (*see also* Residuals)
Authority, centralization and decentralization of, 52–53
Behavior, 6, 8–9, 41–44 (*see also* Values)
Biological indicators, 84, 163
Birds
  diseases, 83
  urban, 83
Birth control, 72–73
Brinkmanship, 148
Butte, Montana (case study), 192–195
Carrying capacity, 22
Change agents, 8
City, a conceptual view, 104–107
Cities, experimental, 78, 109
Climate, and urban growth, 86
Configuration, urban, 87–88, 176
Connectivity, urban, 90–92
  and urban planning, 91

Consumption, 136, 152–153
　reduction of, 126–127, 130–131
Contamination, environmental, 147 (see also Residuals)
Control mechanisms, 143
　functional, 140–141
　structural, 140–141
Communication
　environmental information, 61 (see also Environmental information, Information)
　and technology, 96–97, 110, 111
　and transportation, 96–97
Community
　design/development centers, 57–59, 65
　dialogue, 56, 65
　information exchanges, 56, 65
　information network, 65, 66
　organizations, 56–59
Computer information systems, 189
Crowding, urban, 20 (see also Density)
Cultural institutions, urban, 54–55
Culture, 159
Data
　acquisition, indicators, 154–155
　banks, 13
　dissemination, indicators, 155–156
　processing, indicators, 155
Decision-making
　involvement of public, 7
　processes, 48–49
Density
　patterns, 88–90
　population, 110
　smoothing, 88, 90, 109
　urban, 81, 88–90
Differentiation, urban, 90
Disutility, environmental, 121, 124
Dogs
　diseases, 82, 83
　rabies, 83
　urban, 82–83
　waste disposal problems, 82
Dynamic equilibrium, 38
Ecological accounting systems 59–60, 153–156
Economic growth, 8, 16, 59
Economy, steady state, 38
Ecosystem, 25, 139

Education
　environmental, 43–44, 65, 132–133
　and values, 43
Electrical energy, conservation, 133–134
Energy
　conservation, 133–134, 135, 175
　conservation, and building codes, 133
　conservation, and government, 133
　consumption, rates of, 100, 132
　flows, 11, 19, 25–26 (see also Resources)
　role of, 177
　and transportation, 97–98
Enhancement of human life, 38–40
Environment, contamination of, 147 (see also Residuals)
Environmental agencies, 47
Environmental education, 43–44, 65, 132–133
Environmental ethic, 170
Environmental extension teams, 13–14, 188–189, 191, 203
Environmental impact statements, 62, 116, 159, 176
Environmental information
　communication of, 61, 65, 137, 190–192 (see also Information)
　networks, 13
Environmental monitoring, 60–62 (see also Indicators)
Environmental performance standards, 189–190 (see also Land use standards)
Environmental quality, air, 35–36 (see also Residuals)
Evaluation
　formative, 61–62, 65
　summative, 61–62, 65
Exportation, residuals, 127
Extinction, of species, 147
Family size, 72–73
Federal legislation, environmental, 116–117, 159–160
Feedback mechanisms, 26–28, 29
Fertility, 9, 72–73, 79
Fossil fuels, 100 (see also Resources)
Freedom, systemic, 141
Function, urban, 92–98
Generalization, 149
Goals, 6, 18, 37–41
　and indicators, 169–170
　and models, 166, 167–168

Green space (*see* Open space)
Habitat, 93–94
Habitation
  and technology, 94
  urban, 93–95
Halifax, Nova Scotia (case study), 197
Housing, urban 21, 78, 79 (*see also* Habitation)
Housing subsidies, 78
Impact statements, environmental, 116, 159–160, 176
Importation, resources, 127
Income distribution, 9–10
Indicators
  biological, 84, 163
  aggregation and disaggregation, 157–159
  consistent numerairies, 157
  criteria for selection 156–157
  data acquisition, 153–154
  data dissemination, 155–156
  data processing, 155
  data requirements, 149
  discourse, 139–140
  economic, 60
  environmental, 12, 19, 60–61, 138–163, 176
  exhaustiveness, 156
  and goals, 138–139, 169–170
  gross and refined, 60–61
  inadequacy of, 19
  information generation, 158–159
  independence, 156
  perceptibility, 156
  prices, 159
  relevance, 156
  selection of, 156–157, 174–175
  social and environmental, 19, 138–163
  stability, 161–163
  uses of, 160–161
  resiliency, 161–163
Information, 139
  access to, 49
  communication of, 137, 174
  community exchanges, 56, 65
  community network, 13, 65, 66
  computer systems, 189
  control mechanisms, 143
  hierarchy of levels, 55, 182–189
  use of, 49–50

  and values, 49
Information, environmental
  communication of, 132–133, 203
  exchange of, 182–189
  misuse of, 203
  use of, 190–192
Insects, urban, 82
Intervention, 23, 115–138, 140
  altering product specifications, 126
  classes of, 124–127
  and goals, 118
  decreasing utility, 124
  importing resources and exporting residuals, 124–125
  increasing environmental capacity, 124
  production changes, 125–126
  recycling, 125 (*see also* Recycling)
  reducing consumption, 126–127 (*see also* Consumption)
  resource flows, 120–138
  urban structure and function, 117–120
Institutions, 11–12, 29, 44–59
  collegial organization of, 51
  hierarchical organization of, 50–51
  informal, 18
Institutional change, 50, 54–55
Institutional progress, 44–45
Knoxville, Tennessee (case study), 200–202
Land use
  generative and regenerative, 41
  standards, 62, 109, 176, 186, 180–190
Learning, 45
Legislation, environmental, federal, 116–117, 159–160
Life
  enhancement of, 38–40
  sustaining, 38–40
Mammals, urban, 78 (*see also* Animals)
Market economy, 11–12 (*see also* Economic growth)
McHarg technique, 186
Migration, 9–10, 21, 71–72, 74–76
  age variable, 74–75, 76
  duration of residence, 75
  education variable, 75
  ethnic grouping, 75
  models, 75–76, 110
  patterns, 110–111

Mineral resources, 101 (see also Resources)
Minneapolis-St. Paul, Minnesota (case study), 195–197
Models, 13, 163–173, 175
  change components, 168
  control system on differential equations, 172
  control system with static differential equation, 172
  data display and manipulation techniques, 171–172
  evaluation of, 172–173
  expert team analysis, 172
  feedback control, 171
  and goals, 166, 167–168
  and goals/indicators, 169–170
  large-scale differential equation, 172
  misuse of, 164
  small-scale differential equation, 172
  spatial resolution, 171
  static matrix/differential equation, 172
  static matric technique, 172
  static vs. dynamic characteristics, 171
  and values, 175
Monitoring, environmental, 60–62 (see also Indicators, Land use standards, Performance standards)
Mortality rate, 73–74
New towns, 78, 81
Norman, Oklahoma (model), 202
Natural selection, 142
Open space, 10–11, 21–22, 41–42, 64, 84–85, 94, 95–96, 111, 119–120
  maintenance, 84–85, 110, 119–120
Parks, urban (see Open space)
Participatory planning, 45, 52, 65'
Pest control, 21, 84, 109, 110, 147
Pets, urban 82 (see also Animals)
Performance standards, environmental, 8, 189–190 (see also Land use standards, Monitoring, environmental)
Plants
  allergies, 84
  economic benefits, 84–85
  psychological benefits, 10, 21, 84–85
  urban, 81–85, 110
Planning, 18, 45
Pollution, 23–24 (see also Residuals)
Pollution tax, 176

Population
  age distribution, 21, 79
  animal, 81–85
  birth control, 72–73
  density, 68–70, 110
  distribution trends, 76–81
  ethnic and socio-economic distribution, 79–80
  family size, 72–73
  fertility rates, 67, 72–73
  growth, 16–17, 20, 26–27, 29
  migration, 71–72, 74–76
  mortality rate, 73–74
  negative feedback mechanisms, 70
  non-human urban, 81–85
  plant, 81–85
  pressures, 8–9, 19–20
  rate of growth, 70–76
Prices
  market information, 159
  and pricing, 59–60
Priorities, 45–46
Product specifications, 126
Production changes, 125–126
  effects on space, 129
Production, material, 153
Professionals, urban, 53–54
Rabies, 83
Racism, 10 (see also Population)
Reclamation (see Recycling)
Recreation, 78, 88–90, 95–96, 109
Recycling, 36, 125, 129, 134, 135, 148
  water, 134
Regional boundaries and institutional jurisdiction, 46–49
Regionalization, 48–49, 52
Research
  incentives for, 131, 135
  purpose of, 135
Residential dispersion, 12
Residuals, 101, 167
  discharge, 11, 20, 22–24, 101, 174
  effects on space, 129
  exportation, 124–125, 127
  recycling, 125 (see also Recycling)
  storage, 127–128
  treatment, 124, 127–128, 151–152

Resiliency, 141
  indicators of, 161–163
Resources
  depletion, 11, 100–101, 148, 174
  flows, 19, 98–107
  utilization, historical trend, 39–40
  extraction, 153–154
  fossil fuels, 100
  importation, 124–125, 127
  minerals, 101
  water, 100–101
Responsibility, decentralization of, 52
RUDAT (Regional Urban Design Assistance Team), 190–192
Segregation, 9–10 (*see also* Population, distribution trends, and ethnic and socio-economic distribution)
Sensory overload, 20–21
Site, urban, 86
Social adaptation, 43–44
Social change, 6, 8
Social structures, 51
Space
  effects of production changes on, 129
  effects of residuals discharge on, 129
  utilization of, 101–102, 124
Specialization, 149
Stability, 141
  indicators, 161–163
Steady state economy, 38
Strategic location, urban 86
Structure, urban, 11, 85–92, 117–120
Support systems, natural, 10–11
Survival, 18, 37–38

Sustaining human life, 38–40
System
  definition, 24–25
  models, 13, 163–173, 175
Systems science, 24–25
Systemic freedom
  morphogenic, 141
  morphostatic, 141
Technology, 12–13, 22, 135–136
  and communications, 96–97
  development of, 131
  tinkering, 120
  transportation, 96–97, 110
Transportation, 81, 91, 96–98, 111, 130, 131, 134
  and communications, 96–97
  and energy use, 97–98
  systems, 20, 24
  and technology, 96–97, 110
Unemployment, 71
Urban growth, 7, 17, 20, 22
Urban system, a conceptual view, 104–107
Value conflict, 42
Values, 6–7, 37, 41–44, 137, 143
Waste disposal, landfill, 36
Wastes, (*see* Residuals)
Water, 100–101
  conservation, 133–134
  recycling, 125, 134
Waterloo, Ontario (case study), 197–200
Work, 92–93
  rewards of, 93
Zoning, 8, 62 (*see also* Land use standards, Performance standards)
Zoonoses, 83